The strong principle for the real world is: never use a model if you don't know its limitations and side effects. In fact, you must know what it can't do for you better than what it can do. I am glad this project is taking place: a long-awaited examination of the role—and obligation—of modeling.

Nassim Nicholas Taleb, Distinguished Professor of Risk Engineering, NYU Tandon School of Engineering. Author of the five-volume Incerto series (*The Black Swan*)

The Politics of Modelling: Numbers between Science and Policy is a breath of fresh air and a much-needed cautionary view of the ever-increasing dependence on mathematical modelling in ever-widening directions. The five aspects of modelling that should be 'minded' are a sensitive summary of factors that should be considered when evaluating any mathematical model.

Orrin H. Pilkey, Professor, Duke University's Nicholas School of the Environment. Co-author, with Linda Pilkey-Jarvis, of *Useless Arithmetic: Why Environmental Scientists Can't Predict the Future*, Columbia University Press, Washington, DC, 2009

The methods by which power insinuates itself into models, and facilitates their portability and amendments, are diverse and sometimes insidious. And that's one reason why the range of cases explored in *The Politics of Modelling* are so illuminating and why we need to pay attention to its authors. [...] Good scholarly books usually do one of two things. They dig into the details of something so that we understand it better, see it in a new light. We often call this depth. Or they bring things together in some creative amalgamation that allows us to make new comparisons, to see patterns we hadn't before seen. This we call breadth. It is rare when a book does both things well. This one does.

Wendy N. Espeland, Professor of Sociology, Northwestern University. Author, with Michael Sauder, of *Engines of Anxiety: Academic Rankings, Reputation, and Accountability*, Russell Sage, New York, 2016

A modern Rip Van Winkle, awaking from a century of scientific slumbers, would be dismayed to find so much emphasis on models and so little talk of scientific laws and facts. Although Rip's dyspeptic view of models now seems misguided, a call for caution is very much in order. Modelling tools have consequences both for science and for a larger public, taking in historical, sociological, and moral perspectives as well as technical, scientific ones.

Theodore M. Porter, Department of History, UCLA, author of
***Trust in Numbers*, Princeton University Press,**
Princeton, NJ, 1995

The politics of modelling

The politics of modelling

Numbers between science and policy

Edited by

ANDREA SALTELLI

Universitat Pompeu Fabra (UPF)—Barcelona School of Management

MONICA DI FIORE

National Research Council of Italy (CNR)—Institute for Cognitive Sciences and Technologies

Foreword by

WENDY N. ESPELAND

Preface by

DANIEL SAREWITZ

OXFORD
UNIVERSITY PRESS

OXFORD
UNIVERSITY PRESS

Great Clarendon Street, Oxford, OX2 6DP,
United Kingdom

Oxford University Press is a department of the University of Oxford.
It furthers the University's objective of excellence in research, scholarship,
and education by publishing worldwide. Oxford is a registered trade mark of
Oxford University Press in the UK and in certain other countries

© Oxford University Press 2023

The moral rights of the authors have been asserted

Published in the United States of America by Oxford University Press
198 Madison Avenue, New York, NY 10016, United States of America

British Library Cataloguing in Publication Data

Data available

Library of Congress Control Number: 2023934151

ISBN 978-0-19-887241-2

DOI: 10.1093/oso/9780198872412.001.0001

Printed and bound by
CPI Group (UK) Ltd, Croydon, CR0 4YY

Links to third party websites are provided by Oxford in good faith and
for information only. Oxford disclaims any responsibility for the materials
contained in any third party website referenced in this work.

Foreword

Mathematical modelling as a critical cultural enterprise

Wendy Nelson Espeland[1]
Northwestern University

A few years ago, I was asked to participate in writing a comment for *Nature* about the uses and abuses of modelling.[2] My contributions were meagre, but I was happy to add my name to a list of scholars and practitioners calling for measures to make modelling more socially useful and ethical. The comment took the form of a manifesto, which, according to the dictionaries I checked, is a public pronouncement or declaration that explains or advocates for some vision or policy by a particular group that could be political, artistic, part of a social movement, a school of thought, or anyone wishing to assert a position they consider of public value in an accessible way.[3] A long-winded but plausible definition.

No doubt the most famous manifesto is Karl Marx and Friedrich Engels' 'The Manifesto for the Communist Party', written at the behest of the Party and published in 1848. Perhaps not *quite* as generative (or famous) as that manifesto, the *Nature* comment, 'Five Ways to Ensure that Models Serve Society: A Manifesto', was, nonetheless, an effort to provoke, to change policy, and to offer something of value. It, too, is imbued with a spirit of critique, interrogating the power dimensions of knowledge, and aiming to improve the world. It is also a call to arms. And the critical spirit that informs the manifesto is carried forward in this timely and important book.

The *Nature* manifesto was published in the early days of COVID-19. There is nothing like a global pandemic to bring modelling to the fore. How quickly will the virus spread? Who is likely to get sick? How many people will die or be hospitalized? Is what we are doing to stop the virus working? In the past two-and-a-half years or so, models have been front and centre in our lives as we have struggled over questions like these. *The Politics of Modelling* channels the inspiration of the

---

[1] This brief essay is dedicated to the memory of Art Stinchcombe. He continues to teach and inspire me. Thanks to Sam Carruthers for helpful advice and to Bruce Carruthers for a timely read.

[2] *Nature*, *582*, 25 June 2020.

[3] Oxford English Dictionary, https://www.oed.com; Merriam-Webster Dictionary, https://www.merriam-webster.com/dictionary/manifesto (last accessed 6 April 2023).

manifesto and elaborates its claims. One such claim is that a pandemic is an especially propitious time to consider the role that modelling plays in how we know and experience the world. But that just hints at the book's ambition.

Visibility

Since the early days of the pandemic, we have been inundated with models, and with the statistics, tables, and maps that undergird them and are their products. And many of us have learned bits of the language of diffusion modelling: R, the reproduction rate; transmission coefficients, susceptible individuals; infectives; 'flatten the curve'; stochastic differential equations (OK, that last one is a stretch). Models do many things, but one of the things they do well is to make certain aspects of the world and projections about the world visible in a systematic way. But the visibility afforded by models is complicated. One potential healthy side effect of the pandemic is to focus attention on modelling, to help us notice it. And clearly the authors represented here wish to amplify that particular type of visibility. Scholars of quantification, those who investigate numbers as an 'artefact', the editors' term, agree that to understand the power of numbers, one must first *notice* them and how we use them, to denaturalize them and situate them in the contexts of their production and use.[4] One of the book's many contributions is to do just that: to problematize and concretize modelling, helping those of us who rely on them, often unwittingly, to see them.

Mathematical models organize our lives as well as our thinking. Let's consider that. As the variety of chapters and authors show, models are everywhere; they are indispensable, and they become infused with power. We use models to understand climate change, finance, public policy, risk, disease, crime, inequality, and warfare. Models are a key part of the intellectual infrastructure of health, management, economies, and crisis. They are endlessly adaptable, limited as much by imagination as by evidence. We also use them to understand the flight patterns of Finnish butterflies[5] or feelings of love and hate in a love triangle.[6] Models are fundamental for core features of training in many fields. It's hard now to imagine how we could understand our worlds without models.

And yet. There are dangers inherent in our models, a danger that animates the book's call for 'responsible modelling'. One danger is what models make invisible. This is more than just the cost of simplification, the necessary exclusion that formalization demands. There is also a subtler form of invisibility that is less

---

[4] A few scholars who make this point include: Kruger, Daston, and Heidelberger 1987, Porter 1995, Desrosières 1998, Espeland and Stevens 1998, MacKenzie 2006, Didier 2020, Carruthers 2022, and Mennicken and Salais 2022.

[5] See Harrison, Hanski, and Ovaskainen 2011.

[6] See Sprott 2004.

understood, less publicized than the more obvious rules of formal abstraction. This invisibility or obscurity hinges on the subterranean but difficult to recover assumptions, values, and biases that are quietly built into models. We know, and the chapters show, that models can discriminate, and not in a good way; reinforce geopolitical hegemony; reflect the tensions and trade-offs between sometimes unacknowledged values such as precision and realism, complexity and parsimony, the past, the present, and the future, alternative futures, and the like. For powerful abstractions, models and their interpretations can evoke a range of emotions and attitudes. We may presume their neutrality or become suspicious of unsavoury or self-interested manipulations. We may enjoy the technical pleasures and achievements of abstracting and formalizing. We may judge models to be elegant or ugly. Models often contain untheorized theories of the world, a selective scepticism, politics. Channelling Foucault, the methods by which power insinuates itself into models, facilitates their portability and amendments, are diverse and sometimes insidious. And that's one reason why the range of cases explored in *The Politics of Modelling* are so illuminating and why we need to pay attention to its authors.

Uncertainty

How to incorporate and account for *uncertainty* is a big challenge for responsible modellers. Perhaps even more difficult is the question of how to preserve recognition of that uncertainty once models are inserted into policy and administration. Uncertainly as a problem to be solved or at least tamed has a long and rich history. It troubled Plato, Aristotle, and the Stoics because, while they differed in their conception of uncertainty, they agreed that it made it difficult to create a virtuous society (Hubler 2021). Note the link to ethics.

In social theory, uncertainty is a prominent feature of understanding modernism, power, discourse, and the nature of administration. As Max Weber (1927 (1961), 1978) would have it, uncertainty is the enemy of capitalism and an anathema to bureaucrats. The standardization, documentation, and calculation that he describes as defining features of modernity, of the depersonalization that is the hallmark of bureaucratic rationality, are all largely efforts to manage uncertainty or, to be anachronistic, assess risk.[7] Weber recognized that efforts to tame uncertainty, to control the present and the future, was a utopian quest, one incompatible with the fundamental uncertainties of living. Nonetheless, our drive to keep at it, to keep manufacturing new technologies for doing so, are powerful longings shrouded in potent incentives. And for this deeply ambivalent scholar, the results are decidedly mixed.

---

[7] The classic distinction between uncertainty and risk is Frank Knight's (1921 (2006)).

Michel Foucault (1977, 1991), another critic of modernism, provided a different vocabulary for conceptualizing power. Governmentality, the capacity, and breadth, to control and impose diverse technologies or disciplines of power produces certain kinds of populations and subjects that become capable of policing themselves. We can understand modelling as a practice of governing, one aligned with health, security, consumption, and so on. Assessing risk, no longer anachronistic, is a big target for modellers and is, in this sense, an important form of governmentality. In Michael Power's (2007) apposite phrase, risk is 'organized uncertainty'. The question, as always, is unpacking *how* it is organized and to what effect. And in each of these instances, bureaucratization, governmentality, and risk assessment, efforts to tame uncertainty are reflexive and reactive— knowledge that changes what gets done, who does it, and the relation between the two.

Specialists in the study of organizations use the term 'uncertainty absorption' (March and Simon 1958 (1993): 186–7) to describe the widespread process by which the provisional quality of information is forgotten or 'absorbed' as it travels inside (and outside) of organizations. In their words: 'Uncertainty absorption takes place when inferences are drawn from a body of evidence and the inferences, instead of the evidence itself, are then communicated ... Through the process of uncertainty absorption, the recipient of a communication is severely limited in his ability to judge its correctness'. To extrapolate, the further the information gets from those who make models, those who directly confront their messiness, the easier it is to view the numbers, parameters, variables, and relationships among them as sturdier and more reliable than they are. The nuts and bolts of the inevitable organizational editing that accompanies and informs models can transpose uncertainty into at least a provisional certainty. This makes it easier to fall in love with them. As the editors point out, the hubris of asking more of models than they can reasonably provide, of adding parameters that can overwhelm the evidence, can amplify errors that may go unmarked if the underlying uncertainties of the models are not made evident to users. Or if users choose not to listen.

Perhaps most important is our capacity to forget that the assumptions that undergird them are just that: assumptions. In the case of COVID-19, how many people will we have contact with? How many will wear masks? Who will travel from where? How much isolation will we endure? Assumptions may appear to be static expectations about dynamic social processes. Sometimes assumptions are about the nature and pace of what is dynamic. One danger in modelling is failing to test assumptions and revise them often enough, something that may be hard to do, or over time. I might be willing to wear a mask for a year or to forgo seeing my ailing mother for six months, but my behaviour will likely change as the pandemic persists. We tend to hold fast to our models, to our numbers, and forget all the caveats that we build into them. Sometimes this takes the form of bias error, where there is a difference between a model's predictions and its target value, when compared to the data that trained it. Models can be over-fitted or under-fitted depending on the type and extent of the underlying assumption. We need models that regularly confront the messiness and dynamism of the worlds they depict and are then modified to reflect what is new or what they got wrong.

This requires more than good feedback. It requires the resources, the ethos, and the institutionalized discipline to react in appropriate ways. The challenge of making and adjusting transparent assumptions stems partly from the cost of such vigilance, psychically as well as practically. And as with many of our creations, like children, once they go out into the world, our capacity to control them diminishes. But responsible modelling requires that we 'mind the assumptions', check our 'hubris' about what our models can do, and attend to their consequences. Models are generative and potent; we need to be mindful of what they do and who they affect.

But there is a different problem associated with the uncertainty that models try to express and manage, as this book demonstrates. We often grant too much certainty to their implications or projections. For policy makers, especially in times of crisis, doing something often trumps waiting. And undoing what is done—even when the outcomes or projections of models require revision—is an arduous task, practically, cognitively, and politically. Policy can be a blunt and inflexible instrument; the recursive feedback between the world and the model that we may wish for may be unavailable or intolerable, especially when scientific authority is threatened by polarizing politics. As a recent headline put it: '"Follow the science": As the third year of the pandemic begins, a simple slogan becomes a political weapon.'[8] Sometimes the challenges are more subtle. Educating or swaying public opinion is arduous and expensive, and even positive change can be interpreted as error. *Now* you tell us to wear masks? We no longer need to wash groceries? Six feet isn't far enough apart? Such reversals can amplify public uncertainty and threaten to discredit authority of all kinds, whether scientific, political, or moral. And then, again, there is hubris.

Quantitative authority

Nevertheless, while 'science' may be discredited in certain polarized arenas, more often we grant special authority to quantitative information. Where mere anecdotal information lacks 'objectivity', the 'facticity' of numbers is more taken for granted, sometimes even when they strain credulity. Numbers are hard. Words are soft. Words yield stories rather than 'science'. 'If you can't measure it, you can't manage it.' Sound familiar? And maybe if you can't model it, you can't manage it either. We all know that we can lie with statistics, yet we are still seduced

---

[8] See Fisher 2022.

by the apparent rigour of calculation and mathematics.[9] The long and slippery association of rationality and objectivity with quantification is one reason why even complex calculations and projections are easy to accept. These associations envelop models, too, where they may be even more difficult to deconstruct. If it is often harder, and requires more specialized expertise, to fight with numbers than with words. Models, as a specific type of quantification, may require even more sophisticated contestations. And it is easy to lose face when doing so, which makes it risky.

Models as cultural forms

We can think of models as distinctive cultural forms that deserve their own scrutiny, a scrutiny that extends beyond the technical expertise of those who make and use them. But, like all cultural forms, models make things less visible or invisible as well. Cultural forms include and exclude, according to their specific symbolic boundaries, norms, rules, methods, or genres.[10] If we take models seriously as a cultural artefact, one that is embedded in cultural processes of production, consumption, and reflexivity, then we can deploy cultural theory on the nature of these forms in relation to societies.

One deep thinker in this vein was the German sociologist and critic Georg Simmel (1911 (1968)). In his analysis of modern culture, he described the essential tension and tragedy of modern life as a dialectical relationship between our capacity to produce a dazzling array of cultural forms, and our inability to assimilate them in ways that enrich our lives and cultivate our humanity. Put another way, it is the domination of forms of 'objective culture'—the things we create, such as art, ritual, knowledge, money—over 'subjective culture'—how we use, understand, and assimilate those things or relationships: in his words, our 'ability to embrace, use and feel culture'. One of Simmel's examples is fashion (1904 (1957)). Fashion, an example of objective culture, is a 'form of imitation' that becomes a mode of self-expression. It trickles down, usually from the elite to the lower classes, but once it spreads too broadly to others, the search for new forms of self- expression through clothing becomes paramount. The pressure for new consumption leads to mass production of goods, which quickens the pace of the dialectic and lessens the idiosyncrasy or individuality that can be expressed. The result is an overabundance of clothing that signifies little to those who buy it but who continue to search for authentic self-expression via their clothes. For Simmel, this tragedy is a distinctly modern form of alienation and reification.

---

[9] Sally Merry (2016) shows how these sorts of seductions take place with indicators of human rights, gender violence, and sex trafficking.
[10] See Lamont and Molnár (2002) on the nature of symbolic boundaries.

We can see interesting parallels in the kinds of cultural objects that Simmel studied (most brilliantly in his *Philosophy of Money* (1900 (1978), 2004)) and the mathematical models addressed in this volume. The proliferation of models as a way to understand and administer the world, one made possible by disciplinary developments, computational power, the emergence of global markets in 'big data', artificial intelligence, and so on, proceed at such a pace that we cannot fully embrace or understand them as cultural forms. As forms of performative knowledge, they act on us and on the world in ways we often don't understand or appreciate. Our capacity to absorb them is so limited that they remain externalized, objectified forms for most of us rather than allowing us to cultivate ourselves through them. This tension in contemporary society is endemic, paradoxical, and, for Simmel, inevitably tragic. His analysis of the tragedy of culture has roots in Marx's theories of commodification, Weber's encroaching forms of instrumental and formal rationalization, his own analysis of urbanism and its temporality; and he has influenced subsequent thinkers, including some post-modernist critiques.

Why should we care? Is it so devastating if I don't comprehend how most mathematical models are made or used, or which for which purposes? Can't I just 'trust them' as mere mechanical arbiters of decisions?[11] As the authors make clear, it's not so much that I individually need to understand models and their many consequences. But I am surely better off if I understand their capacity to influence my life, the policies that are made or rejected, the way politics is conducted. I am more in charge of my life and my community if I understand how the knowledge and predictions that are embedded in models are translated into practices that affect me. It is also useful if I can trust that the models are being built and used in appropriate ways by people who are in some sense held accountable, whether this takes the form of scientific review, public deliberation, or voting. More important is that the creators of models devise methods to monitor themselves. Hence the principles outlined in the book.

While Simmel's vision of cultural processes is rooted in the late nineteenth century, it nevertheless remains a powerful and pertinent analysis or, dare I say, model, of one way we can think about developments that are only expanding in scope, sophistication, and pace. What Simmel described and analysed was a relatively new pattern, an emergent and ascending powerful dialectic, one stemming from the profound global changes he witnessed; but he still captures in helpful ways some of the dynamics, or dialectics if you will, of contemporary cultures. For Simmel, we become less human when the relationship between objective and subjective culture is distorted, when we can't keep up. And the result contributes to the kind of tragedy Simmel was suggesting; we risk objectification that we cannot

---

[11] Porter (1995) offers a fine account of how trust in persons has been usurped by trust in numbers in many contexts.

control and becoming subjects who cannot participate in their cultures in the ways that are most meaningful. And those subjects include the purveyor of models.

A way out?

The call to arms in *The Politics of Modelling*, like a good manifesto, like all good criticism, offers a way out, a set of principles, to redress the dangers of modelling. Its contribution builds on work done by sociologists in other contexts. Arthur Stinchcombe (2001), for one, reminds us that it is too facile, too 'romantic', to reject formality outright in favour of that overused term 'lived experience'. His is a qualified defence of formality, but one he believed that scholars and scientists, especially sociologists, needed to heed. If our finely honed, some would say knee-jerk, scepticism of modelling yields only criticism stripped of promise, what have we really accomplished? Sociologists tend to analyse formalization as either 'fraud or ritual'. For Stinchcombe, the merits of formality can be formidable, but they are also deeply contextual and idiosyncratic; he helps unpack the situations when formality mainly 'works'.[12] He conceives of formality as 'abstraction plus government' or governance, and understands its primary function as being 'to bring activities within the purview of reasoning devices—[e.g.] algorithms—that relate abstractions to each other' (ibid.: 38). When social activities are planned or governed with abstract descriptions of what ought to happen—some of his examples are immigration law, standard operating procedures, architectural blueprints—and when these abstract descriptions or kinds of formality are adapted to changing situations, they can be more effective or successful than when we rely on the actual constituents of the relationship or processes at hand. Put another way, planning with formality can be more productive than trying to incorporate too much of the messy complexity and details of the 'real' world.[13]

Stinchcombe offers three variables that matter for discerning when formal abstractions can do a good enough job, when they work best or are able to outperform other less abstract systems. First, useful formality is based on abstractions that are accurate enough depictions of problems and solutions such that, as guides to action, they realize 'cognitive adequacy'. By that Stinchcombe means modellers not only must they avoid 'formalizing ignorance' but must be able to make coherent connections between the system and its outcomes and an evolving world.

---

[12] For a helpful review of Stinchcombe, see Dobbin 2004.

[13] But Stinchcombe's carefully qualified view is far from Milton Friedman's more radical endorsement of 'instrumentalist' assumptions in economics, which he sees as more powerful and predictive than the 'naïve' realist alternative. According to Friedman, 'a theory cannot be tested by comparing its 'assumptions directly with "reality". Indeed, there is no meaningful way in which this can be done' (1953: 41).

Second, communication is key in the sense that outcomes and feedback about outcomes, including errors and changing conditions, are capable of being conveyed in ways that those who make and use the models understand. Correction requires the right forms of communication, and whether it is in the expert language of architectural drafting, contract law, or liquidity markets, models must be comprehensible enough to be transmissible in appropriate forms; be transparent enough to be adjusted; and durable such that the quality of the messages that the models convey does not decay over time. The information embodied in models must be interpretable by the meteorologists, physicists, or lawyers who use them in the contexts in which they use them.

The third criteria for effective formalization is that it contains clear 'trajectories of improvement'. This entails a capacity for correction if, for example, early representations prove wrong, inadequate, or out of date. It also demands robustness such that abstractions 'abstract correctly', not only for the specific purposes for which they were created but also for different situations. In his example, measuring the success of a product line might involve total sales but a more robust measure would be rate of growth or decline in relation to comparable product lines. Another necessary feature of trajectories of improvement is that the informality that is invariably built into or accompanies formalizations either improves the abstraction or improves the embedding of that abstraction. American commercial blood banks, for example, are tightly regulated by Food and Drug Administration rules for protecting donors, and the blood supply relies on standardized indicators of someone being 'healthy enough' to sell their blood products, in this case plasma (Espeland 1984). These formalizations include a range of vital signs like blood pressure, haematocrit, weight, and so on. This works quite well, except in cases where a healthy person somehow doesn't fit the criteria, or an unhealthy one does. Expert informality becomes a way to avoid these errors. For the marathon runner whose pulse and blood pressure were too low, technicians taught him to drink coffee and run to his appointment, guaranteeing that his vitals would be within the proper range. For unhealthy people who met the criteria but lied about their addictions, technicians would produce disqualifying measures. Not strictly legal, perhaps, but these informal interventions were part of a professional ethic of care that undermined formality in the service of good. Without built-in adaptive practices, Stinchcombe argues, formalization will immediately 'kill human creativity', which undermines its value. Getting the informality right matters for effective formality.

Stinchcombe's focus is on particular abstract *systems*, but that unit of analysis works for mathematical modelling as well. Models are social in their creation and use and, as such, they are part of complex systems of formalization that extend from production to reception to adaptations over time. Luckily, the distinctiveness of models as particular forms of formalization, ones that are deployed in diverse institutional domains, is explicitly addressed in *The Politics of Modelling*.

And Stinchcombe's efforts at producing a general theory of formality, and his pre-scriptions for its success, are nicely if not overtly adapted and elaborated by the authors here. Thank you.

At their best, models are helpful, self-conscious simplifications; their criteria of inclusion and exclusion are clear or at least retrievable; and their makers or users do not try to stretch their usefulness or their authority too far. At their worse, their heroic simplifications and obtuseness to changing conditions are harmful in ways that limit their value as representations and guides to action. They induce a complacency, the sense that they are not artefacts, not social or cultural processes rooted in inherited classifications, methods, presumptions, layers of discourse, and particular values and framings expressed as universal. These conditions, this book reminds us, require vigilant mindfulness.

Good scholarly books usually do one of two things. They dig into the details of something so that we understand it better, see it in a new light. We often call this depth. Or they bring things together in some creative amalgamation that allows us to make new comparisons, to see patterns we hadn't before seen. This we call breadth. It is rare when a book does both things well. This one does.

References

Carruthers, B.G. 2022. *The Economy of Promises: Trust, Power, and Credit in America*, Princeton University Press, Princeton.

Desrosières, A. 1998. *The Politics of Large Numbers: A History of Statistical Reasoning*, Cambridge University Press, Cambridge.

Didier, E. 2020. *America by the Numbers: Quantification, Democracy, and the Birth of National Statistics*. The MIT Press, Cambridge, MA.

Dobbin, F. 2004. 'Review of Arthur L. Stinchcombe, When Formality Works: Authority and Abstraction in Law and Organizations'. *American Journal of Sociology*, 109(5), 1244–6.

Espeland, W.N. 1984. 'Blood and Money: Exploiting the Embodied Self'. In *The Existential Self in Society*, edited by J.A. Kotarba and A. Fontana, University of Chicago Press, Chicago, pp. 131–55.

Espeland, W.N. and M. Stevens. 1998. 'Commensuration as a Social Process'. *Annual Review of Sociology*, 24, 313–43.

Fisher, M. '"Follow the Science": As the Third Year of the Pandemic Begins, a Simple Slogan Becomes a Political Weapon'. *The Washington Post*, 11 February 2022. https://www.washingtonpost.com/health/2022/02/11/follow-science-year-3-pandemic-begins-simple-slogan-becomes-political-weapon/

Foucault, M. 1977. *Discipline and Punish: The Birth of the Prison*, Random House, New York.

Foucault, M. 1991. 'Governmentality', translated by R. Braidotti and revised by C. Gordon. In *The Foucault Effect: Studies in Governmentality*, edited by G. Burchell, C. Gordon, and P. Miller, University of Chicago Press, Chicago, IL, pp. 87–104.

Friedman, M. 1953. *Essays in Positive Economics*, University of Chicago Press, Chicago, IL.

Harrison, P.J., I. Hanski, and O. Ovaskainen. 2011. 'Bayesian State-Space Modelling of Metapopulation Dynamics in the Glanville Fritillary Butterfly', *Ecological Monographs*, *81*, 581–98.

Hubler, J.N. 2021. *Overcoming Uncertainty in Ancient Greek Political Philosophy*, Palgrave Macmillan, Cham.

Knight, F. 1921 (2006). *Risk, Uncertainty and Profit*, Dover, Mineola, NY.

Krüger, L., L. Daston, and M. Heidelberger, eds. 1987. *The Probabilistic Revolution Ideas in History*, The MIT Press, Cambridge, MA.

Lamont, M. and V. Molnár. 2002. 'The Study of Boundaries Across the Social Sciences'. *Annual Review of Sociology*, *28*, 167–95.

MacKenzie, D. 2006. *An Engine, not a Camera: How Financial Models Shape Markets*, The MIT Press, Cambridge, MA.

March, J.G. and S. Herbert. 1958 (1993). *Organizations*. Blackwell, Cambridge, MA.

Marx, K. and F. Engels. 1847 (1987). 'The Manifesto for the Communist Party'. *Marxists Internet Archive*. https://www.marxists.org/archive/marx/works/download/pdf/Manifesto.pdf (last accessed 6 April 2023).

Mennicken, A. and R. Salais, eds. 2022. *The New Politics of Numbers: Utopia, Evidence and Democracy*, Executive Politics and Governance, Palgrave Macmillan, Cham.

Merry, S. 2016. *The Seductions of Quantification Measuring Human Rights, Gender Violence, and Sex Trafficking*, University of Chicago Press, Chicago.

Porter, T.M. 1995. *Trust in Numbers*, Princeton University Press, Princeton.

Power, M. 2007. *Organized Uncertainty: Designing A World of Risk Management*, Oxford University Press, Oxford.

Simmel, G. 1904 (1957). 'Fashion'. *American Journal of Sociology*, *62*(6), 541–58.

Simmel, G. 1911 (1968). 'The Concept and Tragedy of Culture'. In *Georg Simmel: The Conflict in Modern Culture and Other Essays*, edited by K. Peter Etzkorn, Columbia Teachers College Press, New York, pp. 27–46.

Simmel, G. 1900 (1978). *The Philosophy of Money*, Routledge, London.

Sprott, J.C. 2004. 'Dynamical Models of Love', *Nonlinear Dynamics, Psychology, and Life Sciences*, *8*(3), 303–13.

Stinchcombe, A. 2001. *When Formality Works: Authority and Abstraction in Law and Organizations*, University of Chicago Press, Chicago.

Weber, M. 1927 (1961). *General Economic History*, Collier Books, New York.

Weber, M. 1978. *Economy and Society*, edited by G. Roth and C. Wittich, University of California Press, Berkeley, CA, pp. 63–206, 956–1006.

Preface

The sciences of modelling through

Daniel Sarewitz

Decisions are forward looking: based on information and preferences in the present, they aim to attain a desired outcome in the future.[1] Whether or not a decision actually achieves its intended outcome will depend in part on whether the information upon which the decision is based is a good indicator of what will happen in the future. Let's say I find a person attractive, I share their values and interests, and I have known them to always act in an honest and honourable way. On this basis, I determine that if I marry the person, I can expect to enjoy a fulfilling, lifetime partnership. The decision depends on a prediction of the desired future outcome, but of course the prediction and the outcome, are uncertain.

It would appear to follow, then, that reliable or certain knowledge of the future would allow for better decisions—that is, decisions better able to achieve their intended outcome—than knowledge based only on present and past conditions from which future outcomes are inferred. Predictions thus afford to those decision makers who possess them a special claim to power. Importantly, for the purposes of this short essay, those who are in the business of making predictions may therefore find their services much desired by decision makers. Such a relationship may itself confer benefits, influence, and power on those who predict. Given that these dynamics flow logically from the nature of decision-making, they have probably always been present in politics. The Old Testament story of Joseph's rise to power in Egypt due to his predictions of future crop failure is an ancient example.

Over the past 50 years or so, expanding scientific and technological capabilities in data collection and numerical modelling have increasingly allowed scientists to make predictions about an array of problems that are of direct interest to policy makers. Understanding when scientific predictions can improve decisions—that is, when they can better allow for decisions that achieve their intended outcomes—ought to be an explicit concern for policy researchers, decision makers, and predictive modellers, but for the most part none of these groups has paid much attention to the problem, perhaps because the case for more predictions seems so clear. It isn't, though.

---

[1] This essay builds on a talk of the same title presented at the University of Oxford Institute for Science, Innovation and Society on 4 March 2019. A podcast of that talk is available at: https://podcasts.ox.ac.uk/science-modelling-through (last accessed 7 April 2023).

Weather 'tis nobler ...

Science-based predictions can bolster accountability in decision-making when they are testable against reality. A familiar case is weather prediction.[2] Numerical weather prediction (NWP) uses ensembles of weather-forecasting models to produce forecasts accurate up to about a week. In part, this accuracy can be achieved because, for the short periods involved, 1) weather systems can be treated as relatively closed, 2) models use real-world measurements (today's conditions) to predict short-term changes, and 3) the results of predictions can be evaluated rigorously. Thus, the accuracy of weather forecasts also reflects a history of improvement over the past decades: millions of forecasts made—and tested against reality—each year allow continual learning from successes and mistakes, and precise measurement of predictive skill.

But that's not all. A sophisticated and diverse enterprise has developed to communicate weather predictions and uncertainties for a variety of users, ranging from the National Oceanic and Atmospheric Administration's Digital Marine Weather Dissemination System for maritime users, to the online hourly forecasts at weather.com, adding value to public data collection. Organizations that communicate weather information understand both the strengths and weaknesses of the predictions, but also the information needs of those using the information.

Meanwhile, people and institutions have innumerable opportunities to apply what they learn from such sources directly to decisions, and to see the outcomes of their decisions—in contexts ranging from scheduling a picnic to scheduling airline traffic. Because people and institutions are continually acting on the basis of weather forecasts, they develop tacit knowledge that allows them to interpret information, accommodate uncertainties, and develop trust based on a shared understanding of benefits. Picnickers, airline executives, farmers, and fishers alike learn how far in advance they should believe forecasts of severe weather in order to make decisions whose stakes range from the relatively trivial to the truly consequential. Even though the modelling outputs often remain inexact and fundamentally uncertain (consider the typical '50% chance of rain this afternoon' forecast), and specific short-term forecasts often turn out to be in error, the continued experience of decision makers with weather predictions allows the predictions to be well integrated into the practice of diverse institutions to achieve desired benefits. We might say that accountability, reliability, and social organization itself emerge from a combination of the predictability of the phenomena, and the good judgement borne from the continued experience of both those making the predictions and those who make use of them.

---

[2] This case, as well as the subsequent discussion of climate sensitivity, is more fully presented in Rayner and Sarewitz (2021).

Honoured in the breach

Macro-economic behaviour, on the other hand, continues to evade efforts at scientific prediction. Nobel Prize winner Paul Romer characterizes the development of macro-economic modelling as a 'regression into pseudoscience' (Romer 2016: 20). In reference to the 2008 economic meltdown, Joseph Stiglitz (2011: 591), another Nobelist, observed: 'The standard macroeconomic models have failed, by all the most important tests of scientific theory. They did not predict that the financial crisis would happen; and when it did, they understated its effects.'

Yet other economists continue to argue for the value of models in economic policy-making. According to Christiano, Eichenbaum, and Trabandt (2018: 133),

> The outcome of any important macroeconomic policy change is the net effect of forces operating on different parts of the economy. A central challenge facing policy makers is how to assess the relative strength of those forces. ... Dynamic stochastic general equilibrium (DSGE) models are the leading tool for making such assessments in an open and transparent manner.

Sbordone and colleagues (2010: 23) similarly state that DSGE models are 'playing an important role in the formulation and communication of monetary policy at many of the world's central banks'.

A look at the decision processes in the US Federal Reserve bank (Fed) explains how these two contradictory sentiments can coexist: the models are indeed used, but the results appear not to be a very important part of the bank's decision processes, which depend instead on the experiences of and argumentation among the Fed's various members, and the collective judgement that emerges. As explained by the current chair of the Fed, '[t]he System's structure encourages exploration of a diverse range of views and promotes a healthy policy debate' (Powell 2017: 10).

One of the Fed's regional bank governors (Fischer 2017: 9) explains further:

> Monetary policy decisions in the United States and elsewhere typically arise from a discussion and vote of a committee ... each [committee] participant brings to the table his or her own perspective or view of the world. Part of their role in these meetings is to articulate that perspective and perhaps persuade their colleagues to revise their own ... These narratives shed light on the real-world developments that lie behind the recorded economic data. ... The information underlying a policy decision is, therefore, crucially shaped by a committee system ... The information includes anecdotes and impressions gleaned from business and other contacts. ... Bringing to the table diverse perspectives is a pragmatic way of confronting such deep sources of uncertainty and deciding how to deal with them.

In democracies, national politicians may find themselves held directly accountable for economic performance, as famously captured in former President Bill Clinton's political mantra, 'It's the economy, stupid.' In the US, this accountability is in part achieved by insulating the Fed's economic decisions from direct political interference, with the goal of assuring that its decisions are responsive to broad national, rather than narrow partisan, interests. In pursuing this responsibility, the role of predictive models at the Fed turns out to be entirely subservient to a pluralistic, deliberative process that amalgamates different sources of information and insight into narratives that help make sense of complex and uncertain phenomena.

The Fed's decision process is not just deliberative but incremental and empirical: the result of deliberations is typically a decision either to do nothing, or to tweak the rate at which the government loans money to banks up or down by a quarter of a per cent, or in times of significant economic stress, such as the recent spike in inflation, by a half a per cent or more. The incremental nature of the decisions allows for feedback and learning, assessed against two policy goals mandated by US Congress: price stability and maximum employment. This incremental, learning-by-doing approach typifies what Charles Lindblom, in his famous 1959 article 'The Science of "Muddling Through"', described as the standard method by which decision makers can most effectively manage complex policy problems. Muddling through, Lindblom (1959: 88) concluded, was 'superior to any other decision making method available for complex problems in many circumstances, certainly superior to a futile attempt at superhuman comprehensiveness'.

What, then, is the role of the DSGE models that aim to predict the 'net effect of forces operating on different parts of the economy'? Because the models have little or no predictive accuracy about important events that can destabilize the economy (just think about COVID-19 or the Russian invasion of Ukraine), their role in the Fed's decision process is apparently totemic. Managing the national economy is something that experts do, and using complicated numerical models is a symbol of that expertise and the authority it confers, inspiring confidence, like the stethoscope around a doctor's neck at bedside, or a priest's vestments. The Fed's real work is one of muddling through.

Signifying nothing

The examples of weather and national economic policies are similar because the consequences of using predictions to make decisions are apparent and inescapable. Decision makers may thus be held accountable for undesired outcomes—and scientists may be blamed for inaccurate decisions (see Pielke 1999, for a detailed case study). In the case of weather, which is predictable in the short term, the result is a complex network of institutions that allow improved decisions in part because of improved foreknowledge. In the case of economic behaviour, which is unpredictable, the result is institutional designs that do not depend on foreknowledge.

In other cases, the links between decisions, outcomes, and accountability are much more attenuated. When the politics are divisive and controversial, the lure of predictive knowledge for decision-making may be both powerful and dangerous. For example, cost–benefit predictions of public works projects like dams and shoreline restoration have long been recognized as highly susceptible to manipulation and corruption aimed at delivering the scientific results sought by decision makers (see, for example, Porter 1996, Pilkey and Pilkey-Jarvis 2007).

Anthropogenic climate change has offered a powerful context for the pursuit of policy-relevant predictive models. In 1991, the climate scientist Jerry Mahlman testified before the US Congress that 'The societal need for accurate and detailed climate change predictions and their effects will increase as fast or faster than the scientific community can provide them' (Committee on Science, Space, and Technology 1991: 35). Subsequent decades have fully vindicated Mahlman's view, but perhaps not with the outcomes he was presuming. In particular, where climate predictions draw close to policy implications, the boundary between science and politics seems to collapse. The key point for the purposes of this essay is that the accuracy of policy-relevant climate predictions is typically impossible to assess, and model performance—predictability—cannot be improved over time, in part because the relevant systems may simply be unpredictable, but also because there is no way to assess improvement, so learning is not possible.

A well-studied instance of these difficulties is 'climate sensitivity', often (but not always) defined as the average atmospheric temperature increase that would occur with an equilibrium doubling of atmospheric carbon dioxide. Climate sensitivity is, in essence, a prediction of how quickly and how much the atmosphere will warm up (given a particular, separate prediction of the future of greenhouse gas emissions). It has come to represent a scientific proxy for the seriousness of climate change, with lower sensitivity signalling less serious or more manageable consequences and higher values meaning a greater potential for catastrophic effects. Narrowing the uncertainty range around climate sensitivity has thus been viewed by scientists as important for informing climate change policies. Yet the number corresponds to nothing real, both in the sense that no such equilibrium state can occur, and also that the idea of a global average temperature is itself a numerical abstraction that collapses a great diversity of temperature conditions (across oceans, continents, and all four seasons) into a single value. What it might mean to narrow uncertainty around such a number is anyone's guess.

Weirdly, though, starting in the 1970s, the numerical representation of climate sensitivity remained constant over four decades—as a temperature range of 1.5 °C to 4.5 °C—even as the amount of science pertaining to the problem has expanded enormously. Starting out as a back-of-the-envelope representation of the range of results produced by different climate models from which no probabilistic inferences could be drawn (Van der Sluijs et al. 1998), climate sensitivity gradually came to represent a probability range around the prediction of future warming.

In 2017, for example, an article by Brown and Caldeira reported an equilibrium climate sensitivity value of 3.7 °C with a 50% confidence range of 3.0 °C to 4.2 °C, while the next year a study by Cox, Huntingford, and Williamson (2018) reported a mean value of 2.8 °C with a 66% confidence range of 2.2 °C to 3.4 °C, and an assessment published in 2020 by Sherwood and a team of 24 other scientists reported a 66% probability range of 2.6 °C to 3.9 °C. The Sherwood team study characterized the initial 1.5 °C to 4.5 °C range, first published in a 1979 National Research Council report, as 'prescient' and 'based on very limited information'. In that case, one might reasonably wonder about the significance of four decades of enormous subsequent scientific effort—the Sherwood paper cites more than 500 relevant studies—leading to, putatively, a slightly more precise characterization of the probability range of a prediction of what is in the first place an abstraction.

How can we understand such an effort? What can we say about the nature and value of the predictions that it produces? Consider another, related example: predictions of the consequences of possible future efforts to cool the atmosphere by increasing the amount of sunlight reflected back into space—solar geoengineering.[3] Echoing Mahlman's 1991 call for predictions, Irvine and colleagues (2017: 93) argue: 'An understanding of the impacts of different scenarios of solar geoengineering deployment will be crucial for informing decisions on whether and how to deploy it ... We suggest that a thorough assessment of the climate impacts of a range of scenarios of solar geoengineering deployment is needed'.

One such effort was the Geoengineering Model Intercomparison Project (GeoMIP), 'whose goal was to understand the robust climate model responses to geoengineering. So far, there have been seven core climate model experiments designed for analyzing the effects of solar irradiance reduction, an increase in the loading of stratospheric sulfate aerosols, and marine cloud (or sky) brightening' (Kravitz et al. 2015: 3380). Use of the word 'experiment' is interesting, as if scientists view numerical models as actual controlled manipulations of something real (see, for example, Lahsen 2005), which they are not. This sort of confusion is common in the scientific description of solar geoengineering models, as in this description by MacMartin and others (2017: 12,574): 'Previous simulations only injected aerosols at a single latitude. We show that if you were to inject aerosols at a combination of multiple different latitudes, you could better tailor the resulting climate response, providing a way of designing solar geoengineering to better meet climate goals'. MacMartin and colleagues go on to explain that:

[3] My understanding of the geoengineering case was greatly enhanced by discussions with Jane Flegal; see Flegal 2018.

The climate effects [of solar geoengineering] depend on the spatial distribution of aerosols within the stratosphere, which is at least partially a design choice enabled by choosing how much mass to inject at multiple different latitudes. Thus the question regarding projected climate effects cannot be fully answered without treating geoengineering as a design problem.

(p. 12,588)

Once model predictions become an input into designing the future conditions of the atmosphere, all sorts of normative claims can be made. For example, Svoboda and colleagues state that solar geoengineering

has the potential to reduce risks of injustice related to anthropogenic emissions of greenhouse gases. Relying on evidence from modeling studies, [solar geoengineering] could have the potential to reduce many of the key physical risks of climate change identified by the Intergovernmental Panel on Climate Change (IPCC). Such risks carry potential injustice because they are often imposed on low-emitters who do not benefit from climate change.

(Svoboda et al. 2019)

Carlson and Trisos, in their 2018 publication, express concern about the possible public health consequences of solar geoengineering, and call for more predictive modelling: 'Given the potential for deployment in the next few decades, we believe the planetary health community faces an urgent need to begin forecasting the shifting burden of disease in an engineered climate' (p. 844). Meanwhile, the DECIMALS Fund (Developing Country Impacts Modelling Analysis for SRM) 'will explore how SRM could affect, amongst other things, dust storms across the Middle East, droughts in Southern Africa, the spread of cholera in South Asia' (DECIMALS Fund 2021).

Solar geoengineering is an entirely speculative technology, one that exists in numerical models, and in some science fiction writing (see, for example, Benford 2014), but not in the real world. In the fake world of models, this technology intervenes to tweak the behaviour of the global atmosphere in particular, predicted ways to achieve particular, desired societal outcomes. In the real world, even in the absence of such interventions, the future of climate, and of society, remains deeply and irreducibly uncertain. Such work seems recognizable as scientific research, and its practitioners stake a claim to legitimacy as such: 'GeoMIP has achieved success on a number of fronts: 15 modeling groups have participated ... GeoMIP has resulted in 23 peer-reviewed publications; and results from GeoMIP were featured in the Fifth Assessment Report of the Intergovernmental Panel on Climate Change' (Kravitz et al. 2015: 3380).

But this work does not add to human knowledge. Actually, it adds to ignorance, because the results of the research have no discernible correspondence to anything real, now or in the future. The research generates, in effect, a sort of knowledge pollution, introduced into the world as if it were real science because it is funded by research agencies, vetted by peer review, cited by international reports. No doubt the scientists who conduct this work are sincere in their efforts, but they labour in a corrupted system that rewards the application of numerical models to fantastical questions to produce quantified nonsense. This state of affairs can exist because, unlike with models that predict the weather or the future of the economy, there are no real-world mechanisms of accountability to provide feedbacks on quality and utility, and there are no decision makers who are actually depending on the predictions to make choices that matter to anyone. Yet the published results, now enshrined in literature as real science, can be harvested by scientists and decision makers alike as if they had knowledge value. Advocates for and opponents of geo-engineering may use the results to advance their positions, but the meaning of the results is entirely political.

One could mention many other more worldly, less whimsical cases of the highly problematic application of predictive models to policy problems, such as nuclear waste disposal, diffusion of renewable energy technologies, COVID-19 management, and ecosystem restoration (see the Supplementary Information accompanying Saltelli et al. 2020), but the geoengineering example stands out as an extreme instance of how the allure and legitimacy of scientific prediction sanitizes even the most fantastical, untethered, untestable assumptions about the world and turns them into quantified statements about supposedly possible futures. It's oracles, goat entrails, and star charts, all the way down.

Dear Brutus

These three brief cases suggest that the social and political aspects of scientific predictions can be explored as functions of two key variables, predictability and accountability, as depicted in the figure below. By accountability I have in mind mechanisms to assure the responsiveness of those making predictions to those who may be affected by the use of the predictions in decision-making.

Fig. 1 is offered as a heuristic, a guide for making inferences about predictive modelling in its scientific and social contexts, and in particular its relation to the achievement of public purposes and values. For example, the high-predictability, low-accountability quadrant might be interpreted as encompassing cases in which the benefits of accurate predictions can be captured by special interests to the detriment of broader public interests. An illustrative instance might be the 'predictive analytics' that are now being deployed for product marketing. Predictive analytics combines advances in large-scale data gathering about consumer behaviour with emerging capabilities in artificial intelligence to 'predict customer behavior and the implications of marketing on it' (Bradlow et al. 2017: 80). Another instance

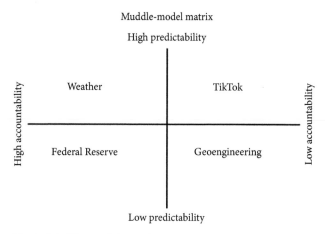

Fig. 1 Muddle-model matrix.

would be the algorithm for the social media app TikTok, which allows user preferences to be predicted with incredible rapidity, so that the app can tailor content offerings in real time that increasingly draw the user in. A third illustration might be the Cambridge Analytica–Facebook scandal associated with the 2016 US presidential election, which points to the potential—if not yet the proven capacity—of using big data analytics to manipulate and accurately predict voter behaviour at large scales. The threats to human agency and political autonomy are real, and make even clearer the links between the ability to predict and the exercise of power.

The relations and distinctions highlighted by the figure can also help to test assumptions and sharpen the insights that underlie public discussions about the role and value of scientific predictions in the political and policy realms. I'll end with three brief examples.

1. The widely reported rise of anti-science, anti-expert sentiments in an increasingly 'post-truth' society (see, for example, d'Ancona 2017) is often associated with those who reject the predictions of climate science. The standard explanation for such sentiments invokes some combination of science illiteracy and misinformation or disinformation, the latter deployed by malign private or political interests to confuse an unsophisticated public. But the 'weather' quadrant allows us to recognize that scepticism about expert predictions is actually a standard, healthy, and practical response to predictive input into decision-making. In Washington, DC, where I live, summer weather forecasts invariably include a prediction of late afternoon thunderstorms, sometimes with very high probability, but anyone who has lived here for a while understands that on many days the storms do not develop, and that cancelling afternoon plans on the basis of weather predictions would mean never having that picnic or going to that baseball game. And anyone who consults the weather app on their telephone quickly comes to understand through

direct experience that forecasts more than a few days into the future are over-precise and often very unreliable. Yet people who might question the accuracy or utility of a given weather forecast are not accused of 'weather science denial'. They are just exercising judgement borne of experience. The predictions provide useful, but not prescriptive, input.

It's thus perhaps paradoxical that those who are sceptical about predictions of future climate behaviour are so often portrayed by those concerned about climate change as anti-science or scientifically illiterate, given that climate predictions carry neither the forecast accuracy nor the understandable implications for decision-making of weather predictions. That one's views on climate predictions have become a political matter is a sign that climate predictions themselves (however scientifically defensible) carry with them significant political content.

2. Indeed, a careful study of Integrated Assessment (IA) models for climate and society by van Beek and colleagues (2022) highlights the centrality of politics to model outputs. IA models are used to generate predictions of alternative futures created by different policy regimes for, say, reducing greenhouse gas emissions. Van Beek et al.'s study documents how the assumptions behind and inputs into the models co-evolve with the changing politics and policy of climate change. The models are 'not neutral "map-makers" but are powerful in shaping the imagined corridor of climate mitigation' (p. 200). For example, as the goal of limiting anthropogenic atmospheric warming to 1.5 °C took hold in policy circles, the assumptions behind and input into IA models had to be changed in order to predict pathways to that desired future. But to do so, modellers had to take heroic steps—starting with the assumption that the 1.5 °C limit was sensible and plausible in the first place, which then created the need to introduce into the models imaginary, fantastical assumptions about technological capabilities (such as nonexistent bioenergy and carbon capture devices) along with the policy and economic frameworks in which they would be deployed (see, for example, Pielke 2018). The consequences here go beyond meaningless predictions. In the absence of any mechanisms of empirical or political accountability (the geoengineering quadrant), the IA models and the politics of climate evolve to reinforce one another in ways that, if translated into actual policy, could greatly narrow the options considered for addressing a complex and uncertain future. This codependency could even lock policy makers into pathways that assure failure, such as depending on imaginary technologies and impossible rates of innovation to achieve policy goals.

3. Which suggests a third problem of predictive science and decision-making: scientific predictions themselves may influence behaviour in ways that make the future even more complex and unpredictable. A powerful example is the risk models (to predict loan default rates) that were used to justify sub-prime mortgage loans by US banks in the years prior to the 2007–8 economic crisis. Such models were used to demonstrate that mortgages that might individually be considered

high risk (so-called 'sub-prime' mortgages) could be grouped together and col-
lectively considered to merit a better (more secure) risk rating. These bundles
of mortgages could then become targets for investors, to be traded like other
securities. The risk models in turn boosted confidence in sub-prime mortgages
themselves, thus feeding back into the system to further inflate housing prices as
more high-risk home buyers were able to get mortgages. As explained by Baily,
Litan, and Johnson (2008),

> computer models took the place of human judgment, as originators did not ade-
> quately assess the risk of borrowers, mortgage services did not adequately assess
> the risk of the terms of mortgage loans they serviced, MBS [mortgage-backed
> securities] issuers did not adequately assess the risk of the securities they sold,
> and so on.

The 2007–8 economic crash had many interlocking causes, one of which was
sub-prime mortgage risk models. But the models also acted to connect and amplify
other causes (for example, lax regulation of the mortgage market, irrational exu-
berance of investors) by concealing behind the apparent logic of a statistical model
the pathological economic behaviour on the part of individuals (for example, people buying homes they couldn't afford) and institutions (for example, banks and
investment firms that didn't understand the risks they were taking on). Economic
actors behaved as if they were in the weather quadrant, but actually they were in
the geoengineering quadrant.

In turn, the subprime mortgage risk models introduced into the economic system new types of human and economic behaviour that were outside the experience
of macro-economic modellers trying to predict the behaviour of national and
international markets. The risk models invisibly helped to remake the economic
future that macro-economic models are supposed to predict. They made the future
even more complex, and created collective ignorance about that complexity. So of
course the macro-economic models both failed to predict the crisis, and greatly
underestimated its magnitude.

'To model or to muddle?' is not quite the question. In the real world, as opposed
to in models, it is the fate of individuals, institutions, and societies to muddle
through their complex and uncertain futures. The impulse to make a hard problem
easier by predictive modelling appears to be a logical consequence of the future-
oriented essence of all decisions. But whether models and the predictions they
produce can guide and improve the procedures for and outcomes of muddling
will strongly depend on specific contexts for decision-making. In some circumstances they can make things worse. When and how to model through seems itself
to be a problem best informed by the sorts of judgement and learning that can only
be acquired by muddling through.

References

Baily, M.N., R.E. Litan, and M.S. Johnson. 2008. 'The Origins of the Financial Crisis'. Brookings Institution Initiative on Business and Public Policy, Fixing Finance Series, Paper 3. https://www.brookings.edu/wp-content/uploads/2016/06/11_origins_crisis_baily_litan.pdf (last accessed 27 July 2022).
van Beek, L., J. Oomen, M. Hajer, P. Pelzer, and D. van Vuuren. 2022. 'Navigating the Political: An Analysis of Political Calibration of Integrated Assessment Modelling in Light of the 1.5° C Goal'. *Environmental Science & Policy, 133*, 193–202.
Benford, G. 2014. 'Eagle'. *Issues in Science and Technology, 30(3)*, 73–83.
Bradlow, E.T., M. Gangwar, P. Kopalle, and S. Voleti. 2017. 'The Role of Big Data and Predictive Analytics in Retailing'. *Journal of Retailing, 93(1)*, 79–95.
Brown, P.T. and K. Caldeira. 2017. 'Greater Future Global Warming Inferred from Earth's Recent Energy Budget'. *Nature, 552(7683)*, 45–50.
Carlson, C.J. and C.H. Trisos. 2018. 'Climate Engineering Needs a Clean Bill of Health'. *Nature Climate Change, 8(10)*, 843–5.
Christiano, L.J., M.S. Eichenbaum, and M. Trabandt. 2018. 'On DSGE Models'. *Journal of Economic Perspectives, 32(3)*, 113–40.
Committee on Science, Space, and Technology, US House of Representatives. 1991. 'Priorities in Global Climate Change Research'. Hearing transcript. US Government Printing Office. ISBN 0-16-037351-4. https://play.google.com/store/books/details?id=8pc8OMGVgYIC&rdid=book-8pc8OMGVgYIC&rdot=1 (last accessed 27 July 2022).
Cox, P.M., C. Huntingford, and M.S. Williamson. 2018. 'Emergent Constraint on Equilibrium Climate Sensitivity from Global Temperature Variability'. *Nature, 553(7688)*, 319–22.
d'Ancona, M. 2017. *Post-Truth: The New War on Truth and How to Fight Back*, Random House, New York.
DECIMALS Fund. 2021. https://www.degrees.ngo/decimals-fund/ (last accessed 27 July 2022).
Fischer, S. 2017. 'Committee Decisions and Monetary Policy Rules'. A speech at 'The Structural Foundations of Monetary Policy', a Hoover Institution Monetary Policy Conference, Stanford University, Stanford, California, 5 May 2017 (No. 951). Board of Governors of the Federal Reserve System (US).
Flegal, J.A. 2018. *The Evidentiary Politics of the Geoengineering Imaginary*, University of California, Berkeley, CA.
Irvine, P.J., B. Kravitz, M.G. Lawrence, and H. Muri. 2016. 'An Overview of the Earth System Science of Solar Geoengineering'. *Wiley Interdisciplinary Reviews: Climate Change, 7(6)*, 815–33.
Irvine, P.J., B. Kravitz, M.G. Lawrence, D. Gerten, C. Caminade, S.N. Gosling, E. Hendy, B. Kassie, W.D. Kissling, H. Muri, A. Oschlies, and S.J. Smith. 2017. 'Towards a Comprehensive Climate Impacts Assessment of Solar Geoengineering'. *Earth's Future, 5(1)*, 93–106, doi: 10.1002/2016EF000389.
Kravitz, B., A. Robock, S. Tilmes, O. Boucher, J.M. English, P.J. Irvine, A. Jones, M.G. Lawrence, M. MacCracken, H. Muri, and J.C. Moore. 2015. 'The Geoengineering Model Intercomparison Project Phase 6 (GeoMIP6): Simulation Design and Preliminary Results'. *Geoscientific Model Development, 8(10)*, 3379–92.
Lahsen, M. 2005. 'Seductive Simulations? Uncertainty Distribution around Climate Models'. *Social Studies of Science, 35(6)*, 895–922.

Lindblom, C.E. 1959. 'The Science of "Muddling Through"'. *Public Administration Review, 19(2)*, 79–88.

MacMartin, D.G., B. Kravitz, S. Tilmes, J.H. Richter, M.J. Mills, J.F. Lamarque, J.J. Tribbia, and F. Vitt. 2017. 'The Climate Response to Stratospheric Aerosol Geoengineering Can Be Tailored Using Multiple Injection Locations'. *Journal of Geophysical Research: Atmospheres, 122(23)*, 12,574–90.

Pielke, R. Jr. 1999. 'Who Decides? Forecasts and Responsibilities'. *Applied Behavioral Science Review, 7(2)*, 83–101.

Pielke, R. Jr. 2018. 'Opening Up the Climate Policy Envelope'. *Issues in Science and Technology 34(4)*, 30–6.

Pilkey, O.H. and L. Pilkey-Jarvis. 2007. *Useless Arithmetic: Why Environmental Scientists Can't Predict the Future*, Columbia University Press, Washington, DC.

Porter, T.M. 1996. *Trust in Numbers*, Princeton University Press, Princeton, NJ.

Powell, J.H. 2017. 'America's Central Bank: The History and Structure of the Federal Reserve'. Remarks at the West Virginia College of Business and Economics Distinguished Speaker Series, Morgantown, WV, 28 March 2017.

Rayner, S. and D. Sarewitz. 2021. 'Policy Making in the Post-Truth World'. *Breakthrough Journal, 13*, 15–43.

Romer, P. 2016. 'The Trouble with Macroeconomics'. Unpublished Working Paper. https://paulromer.net/trouble-with-macroeconomics-update/WP-Trouble.pdf (accessed 27 July 2022).

Saltelli, A., G. Bammer, I. Bruno, E. Charters, M. Di Fiore, E. Didier, W.N. Espeland, J. Kay, S. Lo Piano, D. Mayo, R. Pielke Jr, T. Portaluri, T.M. Porter, A. Puy, I. Rafols, J.R. Ravetz, E. Reinert, D. Sarewitz, P.B. Stark, A. Stirling, J.P. van der Sluijs, and P. Vineis. 2020. 'Five Ways to Ensure That Models Serve Society: A Manifesto'. *Nature, 582*, 482–4.

Sbordone, A.M., A. Tambalotti, K. Rao, and K.J. Walsh. 2010. 'Policy Analysis Using DSGE Models: An Introduction'. *Economic Policy Review, 16(2)*, 23–43.

Sherwood, S.C., M.J. Webb, J.D. Annan, K.C. Armour, P.M. Forster, J.C. Hargreaves, G. Hegerl, S.A. Klein, K.D. Marvel, E.J. Rohling, and M. Watanabe. 2020. 'An Assessment of Earth's Climate Sensitivity Using Multiple Lines of Evidence'. *Reviews of Geophysics, 58(4)*, e2019RG000678, 92p.

Van der Sluijs, J., J. Van Eijndhoven, S. Shackley, and B. Wynne. 1998. 'Anchoring Devices in Science for Policy: The Case of Consensus around Climate Sensitivity'. *Social Studies of Science, 28(2)*, 291–323.

Stiglitz, J.E. 2011. 'Rethinking Macroeconomics: What Failed, and How to Repair It'. *Journal of the European Economic Association, 9(4)*, 591–645.

Svoboda, T. P.J. Irvine, D. Callies, and M. Sugiyama. 2019. 'The Potential for Climate Engineering with Stratospheric Sulfate Aerosol Injections to Reduce Climate Injustice'. *Journal of Global Ethics*, doi: 10.1080/17449626.2018.1552180.

Acknowledgements

Many persons have helped to bring this volume to life.

The curators are especially grateful to Oxford University Press for procuring several precious anonymous reviews that helped improve our work. Special thanks are due to Dan Taber and Giulia Lipparini.

Orrin Pilkey Jr generously offered a text to go in the introduction. Blurbs were offered by him as well as by Wendy N. Espeland, Theodor M. Porter, and Nassim N. Taleb.

As authors, we are grateful to Jerome R. Ravetz and Andy Stirling for revising our own chapters, while the gratitude of Arnald Puy goes to Alba Carmona for her reflections on the concept of 'meta-metaphor' to refer to mathematical models, and that of Philip Stark to Gil Anidjar, Robert Geller, Christian Hennig, Jim Rossi, and Jacob Spertus for helpful suggestions.

Additionally, the authors of chapters 4 and 8 are grateful respectively to the School of Geography, Earth and Environmental Sciences, University of Birmingham, and to the School of Built Environment, University of Reading, for meeting the open access fees for their work.

Andrea Saltelli and Monica Di Fiore, Rome, September 2022

Contents

List of abbreviations

AFRODAD	African Forum and Network on Debt and Development
BSE	Bovine spongiform encephalopathy
CESM	Community Earth System Model
CHAWS	Coupled Human and Water Systems
CI	Composite indicator
CIAO tool	The Composite Indicator Analysis and Optimization Matlab tool
COVID-19	Coronavirus disease 2019
DECIMALS	Developing Country Impacts Modelling Analysis for SRM
DESI	Digital Economic and Society Index
DSGE	Dynamic stochastic general equilibrium
EOQ	Economic Order Quantity
ETI	Energy Transition Index
EU	European Union
FED	Federal Reserve System
GCI	Global Competitiveness Index
GCM	Global Climate Models
GDP	Gross Domestic Product
GENESIS	Generalized Model for Simulating Shoreline Change
GeoMIP	Geoengineering Model Intercomparison Project
GHS	German Historical School of Economics
GIGO	Garbage in Garbage out
GNH	Gross National Happiness
GP	Gaussian process
GR4J	Génie Rural à 4 paramètres Journalier
HDI	Human Development Index
HE	Higher education
IA	Integrated assessment
IAMs	Integrated assessment models
IBJJ	Informational Basis of Judgment in Justice
IBM	International Business Machines
IIASA	International Institute for Applied Systems Analysis
INRAE	Institut national de recherche pour l'agriculture, l'alimentation et l'environnement
IPAT	Impact, Population, Affluence and Technology Impact = Population × Affluence × Technology[1]
IPCC	Intergovernmental Panel on Climate Change

---

[1] These factors enter an equation for the impact of human activity on the environment: Impact = Population × Affluence × Technology.

JANPA	Joint Action on Nutrition and Physical Activity
LPJmL	Lund-Potsdam-Jena managed Land
MATSIRO	Minimal Advanced Treatments of Surface Interaction and Runoff
MBS	Mortgage-backed securities
MERS	Middle East Respiratory Syndrome
MHM	Mesoscale Hydrologic Model
MOW	Model of the World
MPI	Multidimensional Poverty Index
MSY	Maximum Stainable Yield
NASA	National Aeronautics and Space Administration
NATO	North Atlantic Treaty Organization
NCD	Non-communicable diseases
NGOs	Non-governmental organizations
NOAA	National Oceanic and Atmospheric Administration
NPIs	Non-pharmaceutical interventions
NPM	New Public Management
NUSAP	Numeral Unit Spread Assessment Pedigree
NWP	Numerical weather prediction
OECD	Organisation for Economic Co-operation and Development
ORCHIDEE	Organising Carbon and Hydrology in Dynamic Ecosystems
PABE	Public Perceptions of Agricultural Biotechnologies in Europe
PDE	Partial differential equation
PEMM	Political economy of mathematical modelling
PISA	Programme for International Student Assessment
PNS	Post-Normal Science
RRI	Responsible Research and Innovation
SAPEA	Science Advice for Policy by European Academies
SD	Standard deviation
SEIRS	Susceptible-Exposed-Infected-Removed-Susceptible
SIDS	Sudden infant death syndrome
SRM	Solar radiation management
STE	Standard textbook economics
STS	Science and technology studies
THE	Times Higher Education
US	United States
UK	United Kingdom
VSL	Value of a Statistical Life
WTO	World Trade Organization

List of contributors

Wolfgang Drechsler is Professor of Governance at TalTech's Ragnar Nurkse Department of Innovation and Governance, EE, Honorary Professor of University College London in the Institute of Innovation and Public Purpose, UK, and Associate at Harvard University's Davis Center for Russian and Eurasian Studies, US; email: wolfgang.drechsler@taltech.ee.

Emanuele Borgonovo is Full Professor and Head of the Department of Decision Sciences at Bocconi University. He is co-editor-in-chief of the European Journal of Operational Research and co-chair of the Uncertainty Quantification Committee of the European Safety and Reliability Association; email: emanuele.borgonovo@unibocconi.it.

Monica Di Fiore is Researcher at the Institute for Cognitive Sciences and Technologies (ISTC) of the Italian National Research Council, Rome, IT; email: monica.difiore@istc.cnr.it.

Wendy N. Espeland is Professor of sociology at the Weinberg College of Arts and Sciences, Northwestern University. She is author with Michael Sauder of Engines of Anxiety: Academic Rankings, Reputation, and Accountability (2016); email: wendyespeland@gmail.com.

Lukas Fuchs is a postdoctoral research fellow at the Philosophy & Ethics group at Eindhoven University of Technology. He recently received his PhD from the Institute for Innovation and Public Purpose at University College London, UK; email: lukas.fuchs.at@gmail.com.

Marta Kuc-Czarnecka is Adjunct Professor and Deputy Head of Department at the Chair of Statistics and Econometrics, Gdańsk University of Technology, PL. She is vice-head of Rethinking Economics Gdańsk; email: marta.kuc@pg.edu.pl.

Samuele Lo Piano is a Postdoctoral Researcher at the School of the Built Environment, University of Reading, UK; email: s.lopiano@reading.ac.uk.

Arnald Puy is an Associate Professor at the School of Geography, Earth and Environmental Sciences; University of Birmingham, UK; email: a.puy@bham.ac.uk.

Jerome R. Ravetz is Associated with the Institute for Science, Innovation and Society, University of Oxford, UK; email: jerome.ravetz@gmail.com.

Daniel Sarewitz is Emeritus Professor of Science and Society, and was co-director and co-founder of the Consortium for Science, Policy &Outcomes (CSPO) at Arizona State University, Arizona, US; email: Daniel.Sarewitz@asu.edu.

Andrea Saltelli is Academic Counsellor at the Universitat Pompeu Fabra (UPF), Barcelona School of Management, Barcelona, ES; email: andrea.saltelli@bsm.upf.edu.

Luca Savarino is Associate Professor of Moral Philosophy at Dipartimento di Studi Umanistici, Università del Piemonte. Orientale, Vercelli, IT; email: luca.savarino@uniupo.it.

Razi Sheikholeslami is Assistant Professor at the Department of Civil Engineering, Sharif University of Technology, Tehran, Iran; email: razi.sheikholeslami@sharif.edu.

Philip B. Stark is Distinguished Professor of Statistics, University of California, Berkeley, California, US; email: stark@stat.berkeley.edu.

Andy Stirling is Professor of Science and Technology Policy at the Science Policy Research Unit at Sussex University, UK; email: a.c.stirling@sussex.ac.uk.

Paolo Vineis is Chair of Environmental Epidemiology at Imperial College, London, UK, and Head of the Unit of Genetic and Molecular Epidemiology at the Italian Institute for Genomic Medicine (IIGM), Torino, IT; email: p.vineis@imperial.ac.uk.

Ting Xu is Professor of Law at Essex Law School, University of Essex, UK; email: ting.xu@essex.ac.uk.

PART I
MEETING MODELS

1

Introduction

Monica Di Fiore and Andrea Saltelli

The present volume is inspired by a comment published in *Nature* (Saltelli et al. 2020a), in the midst of the COVID-19 pandemic, that suggested a *Manifesto* with five rules for responsible modelling.

Each rule is introduced by a 'Mind the' and followed by five key themes of modelling: the assumptions, the hubris, the framing, the consequences and the unknown.

Why this call for responsible modelling? And for whom is it intended?

The use of the term 'responsible' needs attention. Several definitions of the term are available in relation to responsible science and technology. In the European Union, Responsible Research and Innovation (RRI) is a well-identified frame of science policy. There the goals of the European Commission and the aspiration of scholars of science and technology studies (STS) have at times met and at times clashed (Owen, von Schomberg, and Macnaghten 2021) around the operationalization of the term into six 'keys' (ethics, gender equality, governance, open access, public engagement, and science communication) and related measurable indicators.

Our call for 'responsible modelling' has its roots in the critical reflection on the role of models as forms of knowledge, governance tools (Østebø and Henderson 2020), and in the planning of public decision-making. The *Manifesto* recognized that the quality of a model is a guarantee of its usefulness to society, and that both are earned through a critical process during its construction and use. Its critical five-step process is what the authors agreed to consider responsible modelling, claiming that this is the most effective way to assure its quality.

Defining the quality process of a model is as ambitious as it is vast. The *Manifesto* proposed a synthesis, drawn from a multidisciplinary literature, in the awareness that this definition is a social activity, as we will see shortly, and therefore a process of continuous learning. Modelling combines technical choices with values negotiation; it frames the question and, in doing so, prioritizes an underlying vision.

How might models not be responsible? What could be wrong with them?

We have collected here visions of those critical aspects of modelling from jurists, statisticians, economists, philosophers, and number crunchers. This kaleidoscope

Monica Di Fiore and Andrea Saltelli, *Introduction*. In: *The politics of modelling.* Edited by: Andrea Saltelli and Monica Di Fiore, Oxford University Press. © Oxford University Press (2023). DOI: 10.1093/oso/9780198872412.003.0001

of different ways of seeing is meant to prevent disciplinary tunnel vision or the imposition of a master narrative that neglects a plurality of stakes and concerns in cases where models influence societal perceptions and decisions.

Two major circumstances make the present reflection timely. One is the enhanced production of works of sociology and ethics of quantification (Popp Berman and Hirschman 2018, Mennicken and Espeland 2019), which are not limited to sociologists but include, for example, economists, jurists, data scientists, and philosophers. The other is the increased visibility of mathematical models, specifically those brought about by the COVID-19 epidemic. While models' epistemic authority has increased, so has controversy around their use, with contestations at times reaching the political sphere (Pielke 2020). Academicians are not shy of talking of 'models as public troubles' (Rhodes and Lancaster 2020) or of the 'callous nature of the model based policy' (Hinchliffe 2020).

Furthermore, while ideas of justice, fairness, and ethics are often associated with data, algorithms, and even statistics, they are less often associated with mathematical modelling. Major disciplinary discussions agitate the core of the profession of statistics (Mayo 2018), but hardly touch that of mathematical modelling (Saltelli 2019).

The relative immunity of mathematical models has many causes. Within mathematical modelling, different disciplines are united by the lack of an agreed quality standard (Eker et al. 2018, Padilla et al. 2018). Models have remained relatively less explored even in the work of social scientists. As a result, models are under-explained and overinterpreted. Can the gap be filled? We move in this direction with this book.

Since mathematical models are at the core of this, we need to define what they are. For the Oxford Advanced American Dictionary, a model denotes 'a simple description of a system, used for explaining how something works or for calculating what might happen'.

Models can be used to describe an extensive list of phenomena ranging from population dynamics to consumer behaviour, from resource management to business accounting, from planetary physics to pricing in finance, from educational planning to university rankings, and so on.

More structured definitions are also possible, such as Page's (2018) seven goals of modelling: (1) Reason: To identify conditions and deduce logical implications; (2) Explain: To provide (testable) explanations for empirical phenomena; (3) Design: To choose features of institutions, policies, and rules; (4) Communicate: To relate knowledge and understandings; (5) Act: To guide policy choices and strategic actions; (6) Predict: To make numerical and categorical predictions of future and unknown phenomena; (7) Explore: To investigate possibilities and hypotheticals.

This scheme shows the multiple aims and uses of a model, as well as the plurality of intervening actors. All these possible model goals converge at the idea of a *knowledge* produced on a specific issue, or a *system* in the words of the Oxford entry, providing access to a broad repertoire of techniques.

In the present work, we shall describe a mathematical model as an object that lives its own life, reconstructing what happens when this object is immersed in that system by which or for which it has been created and analysing its performative properties and ethical implications.

Each disciplinary angle offers important lessons about the path a model follows along the boundaries of science, policy, and society. Seen from different scholarly perspectives, the *uncertainty, error,* and *complexity* of the system that models are supposed to describe are important elements of analysis. How are these elements addressed and resolved in the model? Which snapshot of the system is taken, and why? These questions are central to both the production and use of models.

Then there is an aesthetics of modelling (Espeland and Stevens 2008). A sign of inaccurate modelling (or of inaccurate reporting of modelling results) is a profusion of digits conveying the illusion of precision and certainty.

'Expected utility', 'decision theory', 'life cycle assessment', 'ecosystem services', 'sound scientific decisions', and 'evidence-based policy' are at times promoted with a promise of accuracy (closeness to the object modelled) that is implicit in the precision of the measure (its many digits) (Stirling 2019). Yet all these labels pre-suppose a vision of what matters that unavoidably influences or determines the outcome, while offering to the outside an image of neutrality and objectivity. The knowledge asymmetry between producers and users of mathematical modelling (Jakeman, Letcher, and Norton 2006) may easily push to the background nor-mative implications of different modelling choices, and even act to deny the very existence of ignorance. As an example, the modelling of risk linked to the intro-duction of new technologies or products was a constant concern for ecologists, who lamented bias and reductionism (Winner 1989, van Zwanenberg 2020), and the need to 'defog the mathematics of uncertainty' (Funtowicz and Ravetz 1990). How were these digits produced? How can the users make sense of what is offered as a crisp model output? Is the model roughly right or precisely wrong, to use John M. Keynes's words?

Are there cases where we can identify a model as 'bogus'? In an interesting work on the nature of mathematics and its relation to the dream of Descartes of control and domination over nature, mathematicians Philip J. Davis and Reuben Hersh (1986) offer some simple rules for gauging whether an 'application' of mathematics is genuine or outright bogus:

- Does the depth of the real-world problem justify the complexity of the mathematical model?
- Are there any genuine mathematical reasonings or non-trivial calculations carried out which require the resources of the mathematical model being proposed?

- Are the coefficients or parameters in the equations capable of being determined in a meaningful and reasonably accurate way?
- Are the conclusions capable of being tested against real-world data?
- Do any non-obvious practical conclusions follow from the analysis?

In a sense, these questions are still being debated. Models, with their pitfalls, have not lost their capacity to concern, especially when used to make socially relevant decisions. This was noted by economist John Kay in relation to the 'making up' of input data that the model needs but for which no evidence is available (Kay 2012).

It has been argued that modelling is an art (Morrison 1991), or a craft (Rosen 1991), more than it is a science. As with all products of art, models are fragile. Hydrologists realized this fragility several decades ago, when noting that with some 'judicious fiddling' of the input parameters, models could have been made to reach practically every conclusion (Hornberger and Spear 1981). Satire moved in, and a model was imagined which, given the facts and the desired conclusions, could construct a set of 'logically sounding steps' linking the former to the latter (Adams 1987). The point was well expressed in the joke by the mathematician John von Neumann: 'With four parameters I can fit an elephant, and with five I can make him wiggle his trunk' (Dyson 2004).

Reflecting on the same set of apparently timeless questions, econometrician E. Leamer recommended 'shaking' the assumption of a model to make sure that the inference remained solid (Leamer 1985). At the time, this was one of the first recommendations to use sensitivity analysis. Reflecting on the same issue two and a half decades later, Leamer noted that the reason why this was not done systematically by all modellers was that, in its candour, sensitivity analysis exposed the fragility of the inference (Leamer 2010).

Different authors point to models as 'public troubles'(Rhodes and Lancaster 2020), as colonizers of increasing swathes of life (McQuillan 2017), as chameleons opportunistically changing skin colour (Pfleiderer 2020), as responsible for the last recession (Ravetz 2008, Porter 2012, Wilmott and Orrell 2017). They are ubiquitous and indispensable, yet they are rarely—if ever—called to account if something goes wrong.

One motivation of the present work is thus to capture these concerns under multiple perspectives. Other works have preceded us in this venture and with a similar multidisciplinary approach.

This brief excursus shows that questions about the quality of models have been a recurring theme, with a tendency for problems to remain in search of a clear direction of progress. The *Manifesto* aimed to find a clear direction addressing the critical issues discussed thus far to a wider arena, beyond the academy, highlighting the need for a process of reciprocal recognition between models and society, fighting the known asymmetry between developers and users, and reminding modellers of their obligation to transparency and model users of their right to understand.

The breadth of mathematical modelling families and styles, their increasing sophistication and reach, makes the comprehension of these artefacts arduous. If one were to develop a hermeneutics of models—that is, a science of deciphering the content of a model as if it were a religious or philosophical text written in a long-lost language—some key features of models as a bridge between evidence, principles, and inference would need consideration:

- To what extent does a model assume neutrality or declare its normative stance?
- What level of abstraction is adopted by the model?
- How attentive is the model to the charting of uncertainty?

These questions are entirely context dependent: treating all models as if they belonged to Newtonian physics would be unhelpful. Reading the chapters ahead will enable the reader to form an opinion about how these questions should be addressed.

The five rules of the *Manifesto* represent the main thread followed in the volume. Underlying these rules is the practice of sensitivity auditing (Saltelli et al. 2013, Saltelli and Funtowicz 2014, SAPEA 2019), a methodology that allows the full modelling process to be looked at with both quantitative and qualitative tools. This is particularly relevant in the context of impact assessment (SAPEA 2019, European Commission 2021). The five rules aim to guide the reader by keeping these hermeneutic requirements in sight.

The rules and the chapters

We come back to our first question: how do the rules underpin a responsible use of models?

Several scholars of the use of science for policy assume that the choice of the tool determines the outcome of the analysis (Majone 1989, Beck 1992). The first chapter following this preface offers us the perspective of statistics: what is hidden behind a model? **Philip Stark** tells us how this is being debated in statistics.

Then come the five rules, extended and explored in their implications.

Mind the framing. What does it take to understand a model? Do we need to look through the flesh of a model to identify its frame, the bones that constitutes its normative and intellectual support? Several among the authors of the present volume discuss this point, arguing that a clear and transparent exposition of the frames is a prerequisite for responsible use of models. **Mind the hubris.** Once the explicit and implicit assumptions of a modelling exercise have been evaluated, what descriptive power do we ask of a model? What trade-off do we accept as sustainable in terms of certainty versus error, accuracy versus relevance in a model? 'Technical people with training, knowledge, and capabilities like to use their talents to the utmost' (Quade 1980), pushing them to ask the model for a level of detail superior to what the model can stand. When new parameters are added for which insufficient evidence exists, the error of the results may increase, especially in the absence of a rigorous monitoring of uncertainty. And other drawbacks may follow. **Arnald Puy** guides the reader through the epistemological mindset leading to the development of ever-detailed models and discusses some practical dangers derived from this pursuit.

Mind the assumptions. How do we make a model transparent, so as to understand the solidity of its logic?

The technique is never neutral (Saltelli et al. 2020b), and this applies pointedly to models. A cost–benefit analysis and a multi-criteria analysis of the same problem—based, as they are, on different assumptions and methods—will in general produce different results. The same will likely apply when comparing an agent-based model with one written as a set of differential equations, and so on. **Emanuele Borgonovo** tells us about the role of assumptions in a model, and how techniques of uncertainty quantification and sensitivity analysis offer a useful hermeneutics to 'read' into models. Assumptions constitute the very material mathematical models are made of. And yet, what was assumed in? What was assumed out? These may seem trivial questions, except that, over the course of a model's life, momentous assumptions tend to be forgotten and a model crystallizes into a state of apparent neutrality.

Mind the consequences. Once a model comes into action, it has consequences in the system. Since economics is perhaps the field where this impact is the most severe, **Wolfgang Drechsler and Lukas Fuchs** discuss the role of models in governance and economics.

Mind the unknowns. Having arrived this far, what do we still not know about the model? **Andy Stirling** elaborates on the rhetorical use of modelling techniques that 'close down on uncertainty'.

See Box 1.1 for an example of how the rules may impact modelling, in the experience of the practitioner Orrin Pilkey.[1]

---

[1] Orrin H. Pilkey is the James B. Duke Professor of Geology emeritus and Director Emeritus of the Program for the Study of Developed Shorelines at Duke University's Nicholas School of the Environment. He is co-author, with Linda Pilkey-Jarvis, of *Useless Arithmetic: Why Environmental Scientists Can't Predict the Future*, Columbia University Press, Washington, DC, 2009.

Box 1.1 Applying the rules: An example of behaviour of beaches (Courtesy of Orrin H. Pilkey)

Mind the assumptions may be the most important rule to consider when dealing with ever-changing nature. Modelling the behaviour of beaches often involves some basic and wrong assumptions. These include a **Shoreface Profile of Equilibrium,** a uniform sand grain size, no rock outcrops or stumps on the shoreface, and a **closure depth** beyond which no sediment is lost in a seaward direction. The widely used **Bruun Rule** to predict the amount of shoreline retreat caused by sea-level rise in particular is based on poor assumptions.

Mind the unknowns is particularly important for models that predict the timing of beach behaviour in response to coastal engineering. One common beach modelling effort is concerned with the expected life span of nourished beaches, a necessary part of cost–benefit ratio calculations. Based on an examination of more than 2,054 nourished beaches since 1923 in the US, we can see that storms are almost always responsible for the loss of the artificial beach. Of course, no one knows when the next storm will occur, meaning that accurate modelling of beach life spans (and of the timing of most other kinds of beach evolution) is impossible from the start. Global climate change is leading to changes in storm intensity and frequency along with sea-level rises, making beach behaviour projections even more complex.

Mind the consequences is critical, as is best illustrated by the **Maximum Sustainable Yield (MSY)** model used to control the size of fisheries, including the cod catch on the Canadian Grand Banks. By the end of the 1980s, this critical fishery had existed for 500 years and 40,000 Canadians depended on it, directly or indirectly. The fishery was so important to local people that despite the fact that field observations indicated that too many cod were being taken, the fishery continued nonetheless, with the support of the flawed MSY model. It was simply unthinkable that the fishery could fail. But it did. The consequences weren't heeded, and in 1990 the Grand Banks cod fishery collapsed and the fishing fleet was called home.[a]

Mind the framing. The US Army Corps of Engineers is an agency that depends on projects for its survival. Without projects such as beach nourishment and inlet dredging, individual districts would close. For years, the predicted life spans of nourished beaches have been very optimistic. In the Carolinas, nourished beaches last of the order of three years, but life span projections have often been closer to 10 years. The explanation given for the rapid beach loss was that the storm that took away the beach was 'unexpected'.

This optimistic projection of life spans was one factor favouring funding of projects which kept the districts humming along.

ᵃ Note that the classic history of the Grand Banks cod fishery debacle, showing how even the best-intentioned modellers could get it so wrong, is by Alan Christopher Finlayson: *Fishing for Truth a Sociological Analysis: A Sociological Analysis of Northern Cod Stock Assessments from 1977 to 1990*, St John's, Newfoundland, Canada: The Institute of Social and Economic Research, 1994.

It is evident that different responsibilities are activated by the rules. Producers of quantification must not hide or downplay the uncertainty, but should strive to make explicit their normative orientation and actively hunt for underlying or forgotten assumptions in their own work. They should also moderate their ambition of developing larger and larger models, keeping them within the range permitted by the available evidence. Users and communicators need to be critical of the inference received, question the assumptions, mind the consequences of following a model prescription, and accept the existence of ignorance.

Our volume has its list of examples of modelling gone wrong but also a list of good practices which might be adopted to alleviate existing pathologies. So, the remaining five chapters tackle solutions and relevant cases to gauge the impact of these rules.

Samuele Lo Piano and co-authors tell us about the logic of sensitivity auditing and its applications.

Marta Kuc-Czarnecka and Andrea Saltelli guide us into the world of composite indicators, generated by aggregating with different selected rules a set of underlying statistical variables.

How the law reacted to the pandemic based on the evidence from mathematical models is discussed by **Ting Xu**.

The epidemiologist **Paolo Vineis** and the philosopher **Luca Savarino** discuss the use of scientific concepts, observations, and models in their intersection with values in policy-making.

The philosopher **Jerome R. Ravetz** leads us towards 'models as metaphors', showing how these may serve the functioning of an 'extended peer community', as proposed from the school of post-normal science (Funtowicz and Ravetz 1993), within a process of model-based deliberation.

Provocative, but useful. For whom?

We have so far tackled the questions with which we introduced this volume.

Two more questions still remain: why should this volume be useful, and for whom?

Enabling mutual understanding between policy makers, the media, and the lay rank-and-file modeller was one of the objectives of the *Manifesto* and is one theme of the present volume. We believe this is how the model serves society.

As discussed, modelling should experience the same level of introspection and activism that can now be seen at play in other families of quantification, as this will benefit the large audience that modelling has been gathering in recent years.

Some readers will find the content of this work provocative. However, the growing consensus around the need for transparency is pushing in the direction of a mutual understanding between models and society, as advocated in the *Manifesto*.

The ideal readers to whom this volume is addressed are the same as those that the literature has helped us to identify—that is, the actors who are involved in, and influenced by, mathematical modelling. They are modellers who are in active operation—especially those working in the most delicate policy sectors, but also future experts such as graduate students. Both are invited to include in their routines the exercise of critical reflection on modelling, which takes into account not only the technical requirements, but also the broader set of concerns for the quality of models that are addressed in the five rules discussed previously.

The widespread use of mathematical models for decision-making relevant to politics, its social ramifications, and its political impact are at the centre of this book; therefore, the volume also speaks to politicians, regulators, public officials, and the staff of representative political bodies.

Finally, journalists, communicators, and advocates of different causes will find in these chapters clues on how to interpret what happens when thousands of lines of coding become the basis for a decision. An increasingly literate and conflicted audience will increasingly aspire to understand better the politics of mathematical modelling, foremost when these are used to take decisions, be it on health, society, or the environment.

References

Adams, D. 1987. *Dirk Gently's Holistic Detective Agency*, Simon and Schuster, New York.

Beck, P.U. 1992. *Risk Society: Towards a New Modernity*, Sage Publications, Newbury Park, CA (Published in association with Theory, Culture & Society).

Davis, P.J. and R. Hersh. 1986. *Descartes' Dream: The World According to Mathematics*, Penguin Books, London.

Eker, S., E. Rovenskaya, M. Obersteiner, and S. Langan. 2018. 'Practice and Perspectives in the Validation of Resource Management Models'. Nature Communications, 9(1), 5359, doi: 10.1038/s41467-018-07811-9.

Espeland, W.N. and M.L. Stevens. 2008. 'A Sociology of Quantification'. European Journal of Sociology, 49(3), 401–36, doi: 10.1017/S0003975609000150.

European Commission. 2021. Better Regulation: Guidelines and Toolbox. https://ec.europa.eu/info/law/law-making-process/planning-and-proposing-law/better-regulation-why-and-how/better-regulation-guidelines-and-toolbox_en,   Brussels 2021.

Funtowicz, S. and J.R. Ravetz. 1990. Uncertainty and Quality in Science for Policy, Kluwer, Dordrecht, doi: 10.1007/978-94-009-0621-1_3.

Funtowicz, S. and J.R. Ravetz. 1993. 'Science for the Post-Normal Age'. Futures, 25(7), 739–55, doi: 10.1016/0016-3287(93)90022-L.

Hinchliffe, S. 2020. 'Model Evidence – The COVID-19 Case'. Somatosphere, 31 March 2020. http://somatosphere.net/forumpost/model-evidence-covid-19/ (last accessed 20 March 2023).

Hornberger, G.M. and R.C. Spear. 1981. 'An Approach to the Preliminary Analysis of Environmental Systems'. Journal of Environmental Management, 12(1), 6396608.

Jakeman, A.J., R.A. Letcher, and J.P. Norton. 2006. 'Ten Iterative Steps in Development and Evaluation of Environmental Models'. Environmental Modelling & Software, 21(5), 602–14.

Kay, J. 2012. 'Bogus Modelling Discredits Evidence-Based Policy'. The London School of Economics [Preprint]. http://blogs.lse.ac.uk/politicsandpolicy/bogus-modelling-kay/ (last accessed 20 March 2023).

Leamer, E.E. 1985. 'Sensitivity Analyses Would Help'. The American Economic Review, 75(3), 308–13.

Leamer, E.E. 2010. 'Tantalus on the Road to Asymptopia'. Journal of Economic Perspectives, 24(2), 31–46, doi: 10.1257/jep.24.2.31.

Majone, G. 1989. Evidence, Argument, and Persuasion in the Policy Process, Yale University Press, New Haven, CT.

Mayo, D.G. 2018. Statistical Inference as Severe Testing: How to Get Beyond the Statistics Wars, Cambridge University Press, Cambridge, doi: 10.1017/9781107286184.

McQuillan, D. 2017. 'The Anthropocene, Resilience and Post-Colonial Computation'. Resilience, 5(2), 92–109, doi: 10.1080/21693293.2016.1240779.

Mennicken, A. and W.N. Espeland. 2019. 'What's New with Numbers? Sociological Approaches to the Study of Quantification'. Annual Review of Sociology, 45(1), 223–45, doi: 10.1146/annurev-soc-073117-041343.

Morrison, F. 1991. The Art of Modeling Dynamic Systems: Forecasting for Chaos, Randomness, and Determinism, Dover Publications, Mineola, NY.

Østebø, M.T. and R. Henderson. 2020. 'Wrong, but Useful – for Whom? The Politics of Pandemic Modeling'. Somatosphere, 24 June 2022. http://somatosphere.net/2020/wrong-but-useful-modelling.html (last accessed 28 February 2022).

Owen, R., R. von Schomberg, and P. Macnaghten. 2021. 'An Unfinished Journey? Reflections on a Decade of Responsible Research and Innovation'. Journal of Responsible Innovation, 8(2), 217–33, doi: 10.1080/23299460.2021.1948789.

Oxford Advanced American Dictionary. Entry: Model. https://www.oxfordlearnersdictionaries.com/definition/american_english/model_1   (last accessed 28 February 2022).

Padilla, J.J., S.Y. Diallo, C.J. Lynch, and R. Gore. 2018. 'Observations on the Practice and Profession of Modeling and Simulation: A Survey Approach'. *SIMULATION, 94(6)*, 493–506, doi: 10.1177/0037549717737159.

Page, S.E. 2018. *The Model Thinker: What You Need to Know to Make Data Work for You*, 1st edition, Basic Books, New York.

Pfleiderer, P. 2020. 'Chameleons: The Misuse of Theoretical Models in Finance and Economics'. *Economica, 87(345)*, 81–107, doi: https://doi.org/10.1111/ecca.12295.

Pielke, R. Jr. 2020. 'The Mudfight Over "Wild-Ass" Covid Numbers Is Pathological'. Wired, 22 April 2020 [Preprint]. https://www.wired.com/story/the-mudfight-over-wild-ass-covid-numbers-is-pathological (last accessed 20 March 2023).

Popp Berman, E., and D. Hirschman. 2018. 'The Sociology of Quantification: Where Are We Now?' *Contemporary Sociology, 47(3)*, 257–66.

Porter, T.M. 2012. 'Funny Numbers'. *Culture Unbound, 4*, 585–98.

Quade, E.S. 1980. 'Pitfalls in Formulation and Modeling'. In Pitfalls of Analysis, edited by G. Majone and E.S. Quade, International Institute for Applied Systems Analysis, Laxenburg, Austria, pp. 23–43.

Ravetz, J.R. 2008. 'Faith and Reason in the Mathematics of the Credit Crunch'. *The Oxford Magazine. Eight Week, Michaelmas term 14–16* [Preprint].

Rhodes, T. and K. Lancaster. 2020. 'Mathematical Models as Public Troubles in COVID-19 Infection Control: Following the Numbers'. *Health Sociology Review*, 29(2), 177–94, doi: 10.1080/14461242.2020.1764376.

Rosen, R. 1991. *Life Itself: A Comprehensive Inquiry into the Nature, Origin, and Fabrication of Life*, Complexity in Ecological Systems, Columbia University Press, DC.

Saltelli, A., Â. Guimarães Pereira, J.P. van der Sluijs, and S. Funtowicz. 2013. 'What Do I Make of Your Latinorum? Sensitivity Auditing of Mathematical Modelling'. *International Journal of Foresight and Innovation Policy, 9(2/3/4)*, 213–34, doi: 10.1504/IJFIP.2013.058610.

Saltelli, A. 2019. 'Statistical versus Mathematical Modelling: A Short Comment'. *Nature Communications, 10*, 1–3, doi: 10.1038/s41467-019-11865-8.

Saltelli, A., Bammer, G., Bruno, I., Charters, E., Di Fiore, M., Didier, E., Nelson Espeland, W., Kay, J., Lo Piano, S., Mayo, D., Pielke Jr, R., Portaluri, T., M. Porter, T., Puy, A., Rafols, I., Ravetz, J. R., Reinert, E., Sarewitz, D., Stark, P. B., Stirling, A., van der Sluijs, J., and Vineis, P. 2020a. 'Five Ways to Ensure that Models Serve Society: A Manifesto'. *Nature, 582*, 482–4, doi: 10.1038/d41586-020-01812-9.

Saltelli, A., L. Benini, S. Funtowicz, M. Giampietro, M. Kaiser, E. Reinert, J. P. van der Sluis. 2020b. 'The Technique Is Never Neutral: How Methodological Choices Condition the Generation of Narratives for Sustainability'. *Environmental Science and Policy, 106*, 87–98.

Saltelli, A. and S. Funtowicz. 2014. 'When All Models Are Wrong'. *Issues in Science and Technology, 30(2)*, 79–85.

SAPEA, Science Advice for Policy by European Academies. 2019. Making Sense of Science for Policy under Conditions of Complexity and Uncertainty. SAPEA, Berlin. https://www.sapea.info/topics/making-sense-of-science/ (last accessed 20 March 2023).

Stirling, A. 2019. 'How Politics Closes Down Uncertainty', *STEPS Centre*. https://steps-centre.org/blog/how-politics-closes-down-uncertainty/ (last accessed 31 March 2019).

Wilmott, P. and D. Orrell. 2017. *The Money Formula: Dodgy Finance, Pseudo Science, and How Mathematicians Took Over the Markets*, Wiley & Sons, New York.

Winner, L. 1989. *The Whale and the Reactor: A Search for Limits in an Age of High Technology*, University of Chicago Press, Chicago, IL.

van Zwanenberg, P. 2020. 'The Unravelling of Technocratic Orthodoxy'. In *The Politics of Uncertainty*, edited by I. Scoones and A. Stirling, Routledge, Abingdon, pp. 58–72.

2

Pay no attention to the model behind the curtain

Philip B. Stark

Introduction

There are reliable, empirically tested models for some phenomena. These are not the subject of this chapter. There are also many models with little or no scientific connection to what they purport to explain. Such 'ungrounded' models may be used because they are convenient,[1] customary, or familiar—even when the phenomena obviously violate the assumptions of the models.

George Box (1976) famously wrote, 'all models are wrong, but some are useful', a quotation that has been used to justify all kinds of mischief. A better quotation from the same paper is, 'Since all models are wrong the scientist must be alert to what is importantly wrong. It is inappropriate to be concerned about mice when there are tigers abroad.' This chapter is about general issues that make models *importantly wrong*.

For the purpose of prediction, whether a model resembles reality matters less than whether the model's predictions are sufficiently accurate for their intended use. For instance, if a model predicts the fraction of customers who will be induced to buy by a marketing change accurately enough to decide whether to adopt that change, it doesn't matter whether the model is realistic. But for the purposes of explanation and causal inference—including formulating public policy—it is generally important for the map (model) to resemble the real-world territory, even if it does not match it perfectly.

The title of this chapter alludes to the 1939 film *The Wizard of Oz*. When Dorothy and her entourage meet The Wizard, he manifests as a booming voice and a disembodied head amid towering flames, steam, smoke, and coloured lights. During this audience, however, Dorothy's dog Toto tugs back a curtain, revealing The Wizard to be a small, ordinary man at the controls of a machine designed

---

[1] A model may be convenient simply because software to fit or run the model is available. See Stark and Saltelli (2018).

Philip B. Stark, *Pay no attention to the model behind the curtain*. In: *The politics of modelling*. Edited by: Andrea Saltelli and Monica Di Fiore, Oxford University Press. © Oxford University Press (2023). DOI: 10.1093/oso/9780198872412.003.0002

to impress, distract, intimidate, and command authority. The Wizard's amplified, booming voice instructs Dorothy and her entourage to 'pay no attention to the man behind the curtain'.

The man behind the curtain is an apt metaphor for ungrounded models. Impressive computer results and quantitative statements about probability, risk, health consequences, economic consequences, and so on, are often driven by a model behind the curtain—a model that we are discouraged from paying attention to. What appears to be 'science' may be a mechanical amplification of the opinions and ad hoc choices built into the model. The choices in building the model—parametrizations, transformations, assumptions about the generative mechanism for the data, nominal data uncertainties, estimation algorithms, statistical tests, prior probabilities, and other 'researcher degrees of freedom'—are like the levers in The Wizard's machine. And like The Wizard's machine, the models may have the power to persuade, intimidate, and impress, but not the power to predict or control.[2]

This chapter tries to pull back the curtain to reveal how little connection there is between many models and the phenomena they purport to represent. It examines some of the consequences, including the fact that uncertainty estimates based on ungrounded models are typically misleading. Four ideas recur:

1. *Quantifauxcation.* Quantifauxcation is a neologism for the common practice of assigning a meaningless number, then concluding that because the result is quantitative, it must mean something (and that if the number has six digits of precision, they all matter). Quantifauxcation usually involves some combination of data, models, inappropriate use of statistics, and logical lacunae.

2. *Type III errors.* A Type I error is a false positive—that is, one that rejects a true hypothesis. A Type II error is a false negative—that is, one that fails to reject a false hypothesis. A Type III error is to answer the wrong question, for example, to test a statistical hypothesis that has little or no connection to the scientific hypothesis (an example of the package deal fallacy, whereby things that are traditionally grouped together are presumed to have an essential connection) (Stark 2022).

3. *Models and conspiracy theories.* Models and conspiracy theories are technologies for providing a hard-to-falsify explanation of just about anything. Both purport to reveal the underlying causes of complicated phenomena. Both are supported by fallacies, including equivocation, changing the subject (red-herring arguments, ignoratio elenchi), appeals to ignorance (argumentum ad ignorantiam), appeals to inappropriate authority (argumentum ad verecundiam), ignoring qualifications (secundum quid), hasty generalization, package deal, cherry-picking facts, fallacy of the single

---

[2] For a manifesto regarding responsible modelling, see Saltelli et al. (2020), the paper that sparked this book.

cause, faulty analogies, and confirmation bias. 'Science Is Real' used as an unreflective slogan that considers 'scientific' results to be trustworthy and credible—regardless of the quality of the underlying data and the methodology, conflicts of interest, and agenda—is emblematic.

4. *Freedman's Rabbit Theorem:*[3] *You cannot pull a rabbit from a hat unless at least one rabbit has previously been placed in the hat.* Keeping Freedman's Rabbit Theorem in mind can identify when results must depend heavily on assumptions. For instance, if the model inputs are rates but the model outputs are probabilities, the model must assume that something is random to put the probability rabbit into the hat.

For the last 70 years or so, it has been fashionable to assign numbers to things to make them 'scientific'. Qualitative arguments are seen as weak, and many humanities and 'soft' sciences such as history, sociology, and economics have adopted and exalted computation as scientific and objective.[4] An opinion piece in *Nature* (1978) put the issue well:

It is objective and rational to take account of imponderable factors. It is subjective, irrational, and dangerous not to take account of them. As that champion of rationality, the philosopher Bertrand Russell, would have argued, rationality involves the whole and balanced use of human faculty, not a rejection of that fraction of it that cannot be made numerical.

Quantitative arguments and quantitative models require quantitative inputs. In many disciplines, the inputs, procedures, and models are all problematic. We now discuss some specific examples.

Procrustes' quantifauxcation: Forcing incommensurable things to the same scale

In Greek mythology, Procrustes forced travellers to fit his bed, stretching them if they were shorter than the bed and cutting their limbs if they were taller. One common tool for Procrustes' quantification is to combine incommensurable things into an 'index', for example adding or averaging points on Likert scales (Stark and Freishtat 2014), doing arithmetic with the numbers as if they were measurements of the same physical quantity.[5] Cost–benefit analysis (that blurs differences among different kinds of costs and benefits) and quantifying uncertainty using

---

[3] Freedman, personal communication, c.2001. On more than one occasion, he told me, 'To pull a rabbit from a hat, a rabbit must first be placed in the hat.'

[4] For a defence of qualitative reasoning in science, see Freedman (2010b).

[5] See also Desrosières (1998), chapter 3, 'Averages and the Realism of Aggregates'.

'probability' as a catch-all (blurring important differences among sources and types of uncertainty) are examples.

Utility and cost–benefit analyses

It is often claimed that the only rational basis for policy is a quantitative cost–benefit analysis,[6] an approach tied to reductionism as described by Scoones and Stirling (2020) and Stirling in chapter 7 of this volume. But if there is no rational basis for the inputs, how can cost–benefit analysis be rational?

Not only are costs and consequences hard to anticipate, enumerate, or estimate in real-world problems, but behind every cost–benefit analysis is the assumption that all costs and benefits can be put on the same, scale, typically money or 'utility'. As a matter of mathematics, multidimensional spaces are not in general *totally ordered*: a transitive, reflexive, asymmetric relation such as 'has more utility than' does not necessarily hold for every pair of points in the space. The idea that you can rank all aspects of a set of outcomes on the same one-dimensional scale is an *assumption*.

Luce and Tukey (1964) give conditions under which an individual's preferences among items with multiple attributes can be ranked on the same scale. The conditions (axioms) are non-trivial, as we shall see. Consider a collection of items, each of which has two attributes (for example, a cost and a benefit). Each attribute has two or more possible values.

Consider sandwiches, for example (see Table 2.1). Luce and Tukey's *(double) cancellation axiom* requires:

If you prefer peanut butter and jelly to turkey and cranberry sauce, and you prefer turkey and mustard to ham and jelly, then you *must* prefer peanut butter and mustard to ham and cranberry sauce.

For sandwiches, the cancellation axiom does not hold for everyone, because in general, which condiment one prefers depends on the filling, and vice versa.

Table 2.1 Sandwich Attributes

Attribute	Possible attribute values		
Filling	peanut butter	turkey	ham
Condiment	mustard	grape jelly	cranberry sauce

[6] For arguments for and against cost–benefit analysis, see Frank (2000). For a general treatment in the context of policy, see Sassone and Schaffer (1978).

Similar problems arise in more realistic cost–benefit analyses as a result of the Procrustean insistence on putting disparate qualitative dimensions on the same scale.

Failure of the cancellation axiom is not the only issue with cost–benefit analysis. To quantify costs, in effect you must assign a dollar value to human life, including future generations; to environmental degradation; to human culture; to endangered species; and so on. You must believe that scales like 'quality-adjusted life-years' are meaningful, reasonable, and a sound basis for decisions. Insisting on quantifying risk and on quantitative cost–benefit analyses requires doing things that may not make sense technically or morally. Some scientists and philosophers are reluctant to be so draconian (Funtowicz and Ravetz 1994).

Similarly, there is a slogan that *risk equals probability times consequences*. But what if the concept of probability doesn't apply to the phenomenon in question? What if the consequences resist enumeration and quantification, or are incommensurable?

Human preferences are not based on probability times consequences—that is, expected utility. For instance, there is a preference for 'sure things' over bets: many people would prefer to receive $1 million for sure than to have a 10% chance of receiving $20 million, even though the expected return is double (Desrosières 1998: 49). And many are loss-averse: they would prefer a 10% chance of winning $1 million over a 50% chance of winning $2 million with a 50% chance of losing $100,000, even though the expected return of the second wager is 9.5 times larger. In a repeated game, basing choices on expected returns might make sense, but especially in a single play, other considerations may dominate.

Probability

Just as costs and benefits cannot necessarily be put on the same scale, not all uncertainties can be put on the same scale—but many practitioners nonetheless use 'probability' as a catch-all.

Probability has an axiomatic aspect and a philosophical aspect. Kolmogorov's axioms, the mathematical basis of modern probability, are just that: mathematics. *Theories of probability* provide the philosophical glue to connect the mathematics to the empirical world, allowing us to interpret probability statements.[7]

The oldest interpretation of probability, *equally likely outcomes*, arose in the sixteenth and seventeenth centuries in studying games of chance—in particular, dice games (Stigler 1986). This theory says that if a system is symmetric,

---

[7] For a more technical discussion, see Freedman (2010a) and Stark and Freedman (2010); Le Cam (1977). For historical discussions, see Desrosières (1998) and Diaconis and Skyrms (2018).

there is no reason nature should prefer one outcome to another, so all outcomes are 'equally likely'. For instance, if a vigorously tossed die is symmetric and balanced, there is no reason for it to land with any particular side on top. Therefore, all six possible outcomes are deemed 'equally likely', with many consequences following from Kolmogorov's axioms as a matter of mathematics.

This interpretation of probability has trouble with situations that do not have the intrinsic symmetries of coins and dice. For example, suppose you toss a thumb-tack. What is the probability that it lands point up versus point down? There is no obvious symmetry to exploit. Should you simply declare those two outcomes to be equally likely? And how might you use symmetry to make sense of 'the probability of an act of nuclear terrorism in the year 2099'? There are many events for which defining 'probability' using equally likely outcomes is unpersuasive or perverse.

The second approach to defining probability is the *frequency theory*, which defines probability in terms of limiting relative frequencies in repeated trials. According to this theory, 'the chance that a coin lands heads is 1/2' means that if one were to toss the coin again and again, the fraction of tosses that resulted in heads would converge to 1/2. This theory makes particular sense in the context of repeated games of chance, because the long-run frequency that a bet pays off is identified to be its probability, tying 'probability' to a gambler's long-run fortune.

There are many phenomena for which the frequency theory makes sense (for example, games of chance where the mechanism of randomization is known and understood) and many for which it does not. What is the probability that the global average temperature will increase by three degrees in the next 50 years? What is the chance there will be an earthquake with magnitude eight or greater in the San Francisco Bay Area in the next 50 years?[8] Can we keep repeating the next 50 years to see what fraction of the time that happens—even in principle?

The *subjective (or (neo)-Bayesian) theory* may help in such situations. The subjective theory defines probability in terms of degree of belief. According to this theory, 'the chance that a coin lands heads is ½' means that the speaker believes it will land heads with the same strength they believe it will land tails. Probability measures the state of mind of the person making the probability statement.

The theory of equally likely outcomes is about the symmetry of the *coin*. The frequency theory is about what the *coin* will do in repeated tosses. The subjective theory is about what *I believe*. It changes the subject from geometry or physics to psychology. The situation is complicated further by the fact that people are not very good judges of what is going to happen. For making personal decisions—for instance, deciding what to bet one's own money on—the subjective theory may be a workable choice.

---

[8] See Stark and Freedman (2010) for a discussion of the difficulty of *defining* 'the chance of an earthquake'.

Le Cam (1977: 134–5) offers the following observations:

(1) The neo-Bayesian theory makes no difference between 'experiences' and 'experiments'. ...

(4) It does not provide a mathematical formalism in which one person can communicate to another the reasons for his opinions or decisions....

(5) The theory blends in the same barrel all forms of uncertainty and treats them alike.

A fourth approach to making sense of probability models (models that specify probabilities) is to treat them as *empirical commitments* (Freedman and Berk 2010). For instance, coin tosses are not truly random: if you knew exactly the mass distribution of the coin, its initial angular velocity, and its initial linear velocity, you could predict with certainty how a tossed coin will land. But you might prefer to model the toss as random, at least for some purposes. Modelling it as random entails predictions that can be checked against future data—empirical commitments. In the usual model of coin tosses as fair and independent, all 2^n possible sequences of heads and tails in n tosses of a fair coin are equally likely, which induces probability distributions for the lengths of runs, the number of heads, and so on. When the number of tosses is sufficiently large, this model does not fit adequately. In particular, there tends to be serial correlation among the tosses. And the coin is more likely to land with the same side up that started up.[9]

A final interpretation of probability is as *metaphor*. This is how 'probability' is often invoked in policy and science. It asserts that a phenomenon occurs 'as if' it were a casino game. Taleb (2007: 127–9) discusses *the ludic fallacy* of treating all probability as if it arose from casino games:

The casino is the only human venture I know where the probabilities are known, Gaussian (i.e., bell-curve), and almost computable. ... [W]e automatically, spontaneously associate chance with these Platonified games. ... Those who spend too much time with their noses glued to maps will tend to mistake the map for the territory.

Epistemic uncertainty and probability

Many scientists use the word 'probability' to describe anything uncertain. A common taxonomy classifies uncertainties as *aleatory* or *epistemic*. Aleatory uncertainty results from the play of chance mechanisms—the luck of the draw.

---

[9] The chance the coin will land with the same side on top that started on top is estimated to be about 51% by Diaconis, Holmes, and Montgomery (2007). See also https://www.stat.berkeley.edu/%7Ealdous/Real-World/coin_tosses.html, which reports an experiment involving 40,000 tosses (last visited 16 June 2022).

Epistemic uncertainty is 'stuff we don't know' but in principle could learn. Imagine a coin that has an unknown chance p of landing heads. Ignorance of p is epistemic uncertainty. But even if we knew p, we would not know the outcome of the next toss: it would still have aleatory uncertainty.

A standard way to combine aleatory and epistemic uncertainties involves representing epistemic uncertainty as subjective (neo-Bayesian) probability. This erroneously puts individual beliefs on a par with unbiased physical measurements that have known uncertainties—Procrustes' quantifauxcation.

For example, the Intergovernmental Panel on Climate Change (IPCC) treats all uncertainties as random:

> [Q]uantified measures of uncertainty in a finding expressed probabilistically (based on statistical analysis of observations or model results or expert judgment).
> … Depending on the nature of the evidence evaluated, teams have the option to quantify the uncertainty in the finding probabilistically. In most cases, author teams will present either a quantified measure of uncertainty or an assigned level of confidence.[10]

These sentences require people to do things that don't make sense—and that don't work. Subjective probability assessments (even by experts) are generally untethered, and subjective confidence is unrelated to accuracy. Mixing measurement errors with subjective probabilities doesn't work. And climate variables have unknown values, not probability distributions.

Calling two things by the same name does not make them the same. Combining aleatory and epistemic uncertainty by considering both to be 'probabilities' amounts to claiming that there are two equivalent ways to tell how much something weighs: I could weigh it on a scale or I could think hard about how much it weighs. The two are on a par, as if thinking hard about the question produces an unbiased measurement. Moreover, it implies that I know the accuracy of my internal 'measurement'.

Psychology, psychophysics, and psychometrics have shown that people are bad at making even rough qualitative estimates, and that quantitative estimates are usually biased. Moreover, the bias can be manipulated through *anchoring* and *priming*, as described in the seminal work of Tversky and Kahneman (1975). Anchoring, the tendency to stick close to an initial estimate, no matter how that estimate was derived, doesn't just affect individuals—it affects entire disciplines. The Millikan (1913) oil drop experiment to measure the charge of an electron is an example: Millikan's value was too low, supposedly because he used an incorrect

---

[10] Mastrandrea et al. (2010): 2. The authors of the paper have expertise in biology, climatology, physics, economics, ecology, and epidemiology. To my surprise (given how garbled the statistics is), one author is a statistician.

value for the viscosity of air. It took about 60 years for new estimates to climb to the currently accepted value, about 0.8% higher (considerably larger than the error bars). Other examples include the speed of light and the amount of iron in spinach.[11] In these examples and others, it took decades for a discipline to correct the original error, which anchored the estimates that followed.

Tversky and Kahneman also showed that we are poor judges of probability, subject to strong biases from *representativeness* and *availability*, which in turn depend on the retrievability of instances—the vagaries of human memory.[12]

We cannot even judge accurately how much an object in our hands weighs. Direct tactile measurement is biased by the density and shape of the object, and even its colour.[13] The notion that one could just think hard about how global temperature will change in the next 50 years and thereby come up with a meaningful estimate and uncertainty for that estimate is preposterous. Wrapping the estimate with computer simulations distracts, rather than illuminates (Saltelli et al. 2015).

Humans are also bad at judging and creating randomness: we have *apophenia* and *pareidolia*, a tendency to see patterns in randomness.[14] Our attempts to mimic randomness have fewer patterns than genuinely random processes have. For instance, we produce too few runs and repeats (for example, Schulz et al. 2012, Shermer 2008). We are over-confident about our estimates and predictions (for example Kahneman 2011, Taleb 2007; see also 'Explosion of the Uncertainty Space' in chapter 4). And our confidence is unrelated to our actual accuracy (Krug 2007, Chua et al. 2004).[15]

In short, insisting on quantifying all kinds of uncertainty on the same scale—probability—is neither helpful nor sensible.

Rates versus probabilities

It is common to conflate empirical rates with probabilities. The two are quite different. If the 'physics' of the problem does not put the probability 'rabbit' into the hat, a model cannot magically produce a probability from a rate.

---

[11] It is widely believed that spinach has substantially more iron than other green vegetables. This is evidently the result of a transcription error in the 1870s that shifted the decimal, multiplying the measured value by 10 (see, e.g., http://www.dailymail.co.uk/sciencetech/article-2354580/Popeyes-legendary-love-spinach-actually-misplaced-decimal-point.html0, last accessed 17 March 2023). That the original value was far too high was well known before the Popeye character became popular in the 1930s.

[12] Their work also shows that probability judgements are insensitive to prior probabilities and to predictability, and that people ignore regression to the mean—even people who have had formal training in probability. (Regression to the mean is the mathematical phenomenon that in a sequence of independent realizations of a random variable, extreme values are likely to be followed by values that are closer to the mean.)

[13] E.g. Bicchi et al. (2008: section 4.4.3).

[14] https://en.wikipedia.org/wiki/Apophenia,   https://en.wikipedia.org/wiki/Pareidolia   (last accessed 16 November 2016).

[15] E.g. https://en.wikipedia.org/wiki/Overconfidence_effect (last accessed 16 November 2016).

Everyday language does not distinguish between 'random', 'haphazard', and 'unpredictable', but the distinction is crucial for scientific inference. 'Random' is a very precise term of art.

Here is an analogy: to tell whether a pot of soup is too salty, a good approach is to stir the soup thoroughly, dip in a tablespoon, and taste the contents of the tablespoon. That amounts to tasting a random sample of soup. If instead of stirring the soup I just dipped the spoon in without looking, that would be a *haphazard* sample. It may be *unpredictable*, but it is not a random sample of soup in the technical sense, and it is not possible to usefully quantify the uncertainty in estimating the saltiness of the pot of soup from a haphazard sample.

Probability, *P*-values, confidence intervals, and so on are meaningful only if the data have a random component, for example if they are a random sample, if they result from random assignment of subjects to different treatment conditions, or if they have random measurement error. They do not make sense for samples of convenience or haphazard samples, and they do not apply to populations. The mean and standard deviation of the results of a group of studies or models that is not a sample from anything does not yield actual *P*-values, confidence intervals, standard errors, and so on.

Rates and probabilities are not the same, and ignorance and randomness are not the same. Not all uncertainties can be put on the same scale.

Probability models in science

How does probability enter a scientific problem? It could be that the underlying physical phenomenon is random, as radioactive decay and other quantum processes are, according to quantum mechanics. Or the investigator might create randomness, for example by conducting a randomized experiment or by drawing a random sample.

Probability may enter as a (*subjective*) *prior probability* to represent epistemic uncertainty, as described above. Under some conditions, the data 'swamp' the prior as more and more observations are made, so the subjectivity of the prior is (eventually) not important. However, the conditions are delicate (see, for example, Diaconis and Freedman 1986, Freedman 1999). Nor is it true that nominally 'uninformative' (that is, uniform) prior distributions are uninformative (Backus 1987, Stark and Tenorio 2010, Stark 2015).

Beyond the technical difficulties, there are practical issues in eliciting subjective distributions, even in simple problems.[16] In fact, priors are rarely elicited—instead, priors are chosen for mathematical convenience or from habit. Some proponents of the neo-Bayesian approach argue that if you are forced to cover all possible bets

---

[16] See, e.g., O'Hagan (1998). Typically, some functional form is posited for the distribution, and only some parameters of that distribution, such as the mean and variance or a few percentiles, are elicited.

and you do not bet according to a prior, there are collections of bets where you are guaranteed to lose money, no matter what happens—so you are not rational if you are not Bayesian. But of course, one is not forced to place bets on all possible outcomes (Freedman 2010a). See also the section named 'Probability', above.

Yet another way probability can enter a scientific problem is through the invention of a *probability model* that is supposed to describe a phenomenon, for example, a regression model, a Gaussian process model, or a stochastic partial differential equation (PDE). But in what sense does the model describe reality, and how well?[17] Describing data compactly (in effect, fitting a model as a means of data compression), predicting what a system will do next, and predicting what a system will do in response to a particular intervention are very different goals. The last involves causal inference, which is far more difficult than the first two (Freedman 2010c).

Fitting a model that has a parameter called 'probability' to data does not mean that the estimated value of that parameter estimates the probability of anything in the real world. Just as the map is not the territory, the model is not the phenomenon, and calling something 'probability' does not make it a probability.

Finally, probability can enter a scientific problem as *metaphor*: a claim that the phenomenon in question behaves 'as if' it is random. What 'as if' means is rarely made precise, but this approach is common, for example in models of earthquakes.[18]

Creating randomness by taking a random sample or assigning subjects at random to experimental conditions is quite different from *inventing* a probability model or proposing a metaphor. The first may allow inferences if the analysis properly accounts for the randomization, if there are adequate controls, and if the study population adequately matches the population about which inferences are sought. But if the probability exists only within an invented model or as a metaphor, the inferences have little foundation. The assumptions and the sensitivity of the conclusions to violations of the assumptions have to be checked in each application and for each set of data: one cannot 'borrow strength' from the fact that a model worked in one context to conclude that it will work in another context.

Simulation and probability

There is a popular impression in several modelling communities that probabilities can be estimated in a 'neutral', assumption-free way by Monte Carlo simulation: just let the computer generate the distribution.

---

[17] These problems may be worse in numerical modelling than in statistical modelling, yet they have received less systematic attention (Saltelli 2019).

[18] See, e.g., Stein and Stein (2013), who claim that the occurrence of earthquakes and other natural hazards is like drawing balls from an urn, and discuss whether and how the number of balls of each type changes between draws.

For instance, an investigator might posit a numerical model for some phenomenon. The values of some parameters in the model are unknown. In one approach to uncertainty quantification, values of those parameters are drawn pseudo-randomly from an assumed joint distribution (generally treating the parameters as independent). The distribution of outputs is often interpreted as the probability of outcomes in the real world.

Monte Carlo simulation is not a way to *discover* the probability distribution of anything; it is a way to estimate the numerical consequences of an *assumed* distribution. It is a substitute for doing an integral or a sum, not a way to uncover the laws of nature. Monte Carlo doesn't tell you anything that wasn't already baked into the simulation. The distribution of the output comes from assumptions in the input (modulo bugs): the probability model for the parameters that govern the simulation. It comes from the computer program, not from the real world. Monte Carlo merely reveals consequences of the assumptions. The randomness is an assumption. The rabbit goes into the hat when you build the probability model and write the software.

Similarly, Gaussian process (GP) models (Kennedy and O'Hagan 2001) are common in uncertainty quantification. I have yet to see a situation outside physics where there was a scientific basis to think that the unknown is approximated well by a GP. The computed uncertainties based on GP models do not mean much if the phenomenon is not (approximately, in a suitable sense) a realization of a GP.

This is not to impugn uncertainty quantification and global sensitivity analysis—both of which can be vital and should be routine—but to clarify that Monte Carlo studies start by *assuming* a probability distribution for various inputs, then simulate the distribution of various outputs. The output distribution may have little connection to real-world uncertainty of the outputs, especially if the input distributions are simply made up, for example as an attempt to represent epistemic uncertainty as a prior distribution, as described above (Stark and Tenorio 2010, Stark 2015). As a rule of thumb, in such circumstances the output uncertainties are at best a lower bound on the true uncertainties.

Cargo-cult confidence

Suppose you have a collection of numbers, for example a multi-model ensemble of climate predictions for global warming,[19] or a list of species extinction rates for some cluster of previous studies. Take the mean (m) and the standard deviation (sd) of this list of numbers. Report the mean as an estimate of something. Calculate the interval [m−1.96sd, m+1.96sd]. Claim that this is a 95% confidence interval or that there is a 95% chance that this interval contains 'the truth'. *If* the collection

---

[19] This is essentially what IPCC does.

of numbers were a random sample from some population with finite variance and *if* that population had a Gaussian distribution (or if the sample size were large enough), *then* the interval would be an approximate confidence interval—but the probability statement would still be gibberish.[20]

But if the list is not a random sample or a collection of measurements with random errors, there is nothing stochastic in the data, and hence there can be no actual confidence interval. Calculating confidence intervals from 'found' data is quantifauxcation (see, for example, van der Sluijs 2016).

A 95% confidence interval for a parameter is an interval calculated from a random sample using a method that has at least a 95% probability producing an interval that includes the true value of the parameter. Once the sample has been drawn, the computed interval either does or does not include the true value of the parameter: there is nothing random about it.

Often, people act as if the true parameter value were random and followed a probability distribution whose expected value is the estimate. The claim is backwards: if the estimate is unbiased, the *estimate* is random, with a probability distribution centred at the true, fixed parameter value—not vice versa. Once the estimate has been made, it is also a fixed quantity, not random.

Cargo-cult confidence is common in IPCC reports. For instance, the 'multi-model ensemble approach' involves computing the mean and standard deviation of a group of model predictions, treating the mean as if it were the expected value of the output and the standard deviation as if it were the standard error of the climate variable.[21]

Avian–turbine interactions

Wind turbine generators occasionally kill birds—in particular, raptors (Watson et al. 2018). This leads to a variety of questions. How many birds, and of what species? What design and siting features of the turbines matter? Can you design turbines or wind farms in such a way as to reduce avian mortality? What design changes would help?

I was peripherally involved in this issue for the Altamont Pass wind farm in the San Francisco Bay Area. Raptors are rare; raptor collisions with wind turbines are rarer. To measure avian mortality from turbines, people look for pieces of birds underneath the turbines. The data aren't perfect. There is background mortality

---

[20] See Stark and Saltelli (2018). Cargo-cult confidence intervals involve the same calculations as confidence intervals, but they are missing a crucial element: the data are not a random sample.

[21] The IPCC also talks about simulation errors being 'independent', which presupposes that such errors are random: https:/archive.ipcc.ch/publications_and_data/ar4/wg1/en/ch10s10-5-4-1.html (last accessed 24 March 2023). But modelling errors are *not* random—they are a textbook example of systematic error. And even if they were random and independent, averaging would tend to reduce the variance of the result, but not necessarily improve accuracy, since accuracy depends on bias as well.

unrelated to wind turbines. Generally, you find bird fragments, not whole birds. Is this two pieces of one bird or pieces of two birds? Carcasses decompose. Scavengers scavenge. Birds may land some distance from the turbine they hit. How do you figure out which turbine is the culprit? Is it possible to make an unbiased estimate of the number of raptors killed by the turbines? Is it possible to relate mortality to turbine design and siting reliably?

The management of the Altamont Pass wind farm hired a consultant, who modelled collisions using a zero-inflated Poisson distribution[22] with parameters that depend on selected properties of the turbines. According to the model, collisions are random and independent, with a probability distribution that is the same for all birds. The expected rates follow a hierarchical Bayesian model that relates them parametrically to properties of the location and design of the turbine.

According to this model, when a bird approaches a turbine, it tosses a biased coin. If the coin lands heads, the bird hits the turbine. If the coin lands tails, the bird avoids the turbine. The chance the coin lands heads depends on the location and design of the turbine, according to a prespecified formula that depends on the turbine location and specific aspects of its design. For each turbine location and design, every bird uses a coin with the same chance of heads, and the birds all toss the coin independently.

But why are collisions random? Why independent? (Birds, even raptors, fly in groups.) Why Poisson? Why do all birds use the same coin for the same turbine, regardless of their species, the weather, windspeed, and other factors? Why doesn't the chance of detecting a bird depend on how big the (piece of) bird is, how tall the grass is, or how long it's been since the last survey?

The analysis is a red herring. It changes the subject from 'How many birds does this turbine kill?' to 'What are the values of coefficients in this zero-inflated hierarchical Bayes Poisson regression model?' It is a Type III error: testing a statistical model with little connection to the scientific question. Rayner (2012: 120) refers to this as *displacement*:

Displacement is the term that I use to describe the process by which an object or activity, such as a computer model, designed to inform management of a real-world phenomenon actually becomes the object of management. Displacement is more subtle than diversion in that it does not merely distract attention away from an area that might otherwise generate uncomfortable knowledge by pointing in another direction, which is the mechanism of distraction, but substitutes a more manageable surrogate. The inspiration for recognizing displacement can be traced to A. N. Whitehead's fallacy of misplaced concreteness, 'the accidental error of mistaking the abstract for the concrete.'

---

[22] This is a mixture of a point mass at zero and a Poisson distribution.

The Rhodium Group American Climate Prospectus

The Bloomberg Philanthropies, the Office of Hank Paulson, the Rockefeller Family Fund, the Skoll Global Threats Fund, and the TomKat Charitable Trust funded a study[23] that purports to predict various impacts of climate change.

The report is somewhat circumspect at the start:

> While our understanding of climate change has improved dramatically in recent years, predicting the severity and timing of future impacts is a challenge. Uncertainty around the level of greenhouse gas emissions going forward and the sensitivity of the climate system to those emissions makes it difficult to know exactly how much warming will occur and when. Tipping points, beyond which abrupt and irreversible changes to the climate occur, could exist. Due to the complexity of the Earth's climate system, we don't know exactly how changes in global average temperatures will manifest at a regional level. There is considerable uncertainty.

But then the report makes rather bold claims:

> In this climate prospectus, we aim to provide decision-makers in business and government with the facts about the economic risks and opportunities climate change poses in the United States.

Yep, the 'facts'. They estimate the effect that climate change will have on mortality, crop yields, energy use, the labour force, and crime, *at the level of individual counties in the United States until the end of the year 2099*. They claim to use an 'evidence-based approach'.[24]

Among other things, the prospectus predicts that violent crime will increase just about everywhere, with different increases in different counties. How do they know? In some places, on hot days there is on average more crime than on cool days.[25] Fit a regression model to the increase. Assume that the fitted regression model is a *response schedule*: it is how nature generates crime rates from temperature. Input the average temperature change predicted by the climate model in each county; out comes the average increase in crime rate.

---

[23] Houser et al. (2015). See also http://riskybusiness.org/site/assets/uploads/2015/09/RiskyBusiness_Report_WEB_09_08_14.pdf (last accessed 16 November 2016).

[24] To my eye, their approach is 'evidence-based' in the same sense that alien abduction movies are 'based on a true story'.

[25] Ranson (2014) claims, 'Between 2010 and 2099, climate change will cause an additional 22,000 murders, 180,000 cases of rape, 1.2 million aggravated assaults, 2.3 million simple assaults, 260,000 robberies, 1.3 million burglaries, 2.2 million cases of larceny, and 580,000 cases of vehicle theft in the United States.'

The hubris of these predictions is stunning. Even if you knew exactly what the temperature and humidity would be in every cubic centimetre of the atmosphere every millisecond of every day, you would have no idea how that would affect the crime rate in the US next year, much less in 2099, much less at the level of individual counties. And that is before factoring in the uncertainty in climate models, which is enormous for global averages (Regier and Stark 2015), and higher for smaller areas, such as individual counties. And it is also before considering that society is not a constant: changes in technology, wealth, and culture over the next 100 years surely matter, as evidenced by the recent rise in mass shootings (Ogasa 2022).

Climate models are theorists' tools, not policy tools: they might help us understand climate processes, but they are not a suitable basis for planning, economic decisions, and so on (see, for example, Saltelli et al. 2015). The fact that intelligent, wealthy people spent a lot of money to conduct this study and that the study received high-profile coverage in *The New York Times* and other visible periodicals shows how effective quantifauxcation is rhetorically.

Conclusion

Uncritical reliance on models has much in common with belief in conspiracy theories. Models and conspiracy theories purport to reveal the underlying causes of complicated phenomena. Both are well suited to reinforcing pre-existing beliefs through confirmation bias.

Proponents of models and conspiracy theories generally treat agreement with a selected set of facts as affirmative evidence, even when there are countless competing explanations that fit the evidence just as well—and evidence that conflicts with their story.

Proponents generally look for confirmation rather than for alternative explanations or critical experiments that could 'stress test' their position (Mayo 2018). When they do face inconvenient facts, modellers and conspiracy theorists tend to complicate their stories—for example, by adding parameters to a model or agents to a conspiracy theory—rather than reconsider their approach.

Both often rely on authority or tribalism for support. For instance, there are entire disciplines built around a particular model, as econometrics is largely built on linear regression and seismic engineering is largely built on probabilistic seismic hazard analysis.

We have discussed several applications where models may conflict with reality in ways that may be obscure to model users, for example because the model forces all costs and benefits to the same scale or treats all uncertainties as probabilities.

The next five chapters describe the principles from the work that sparked this volume (Saltelli et al. 2020), to help ensure that models serve society:

- Mind the framing: Match purpose and context
- Mind the hubris: Complexity can misfire
- Mind the assumptions: Quantify uncertainty and assess sensitivity
- Mind the consequences: Quantification in economic and public policy
- Mind the unknowns: Exploring the politics of ignorance in mathematical models

Acknowledgements

I am grateful to Gil Anidjar, Monica Di Fiore, Robert Geller, Christian Hennig, Jim Rossi, Andrea Saltelli, and Jacob Spertus for helpful suggestions.

References

Backus, G.E. 1987. 'Isotropic Probability Measures in Infinite-Dimensional Spaces'. *Proceedings of the National Academy of Science*, 84, 8755–7.

Bicchi, A., M. Buss, M.O. Ernst, and A. Peer. 2008. *The Sense of Touch and Its Rendering: Progress in Haptics Research*, Springer-Verlag, Berlin and Heidelberg.

Box, G.E.P. 1976. 'Science and Statistics'. *Journal of the American Statistical Association*, 71, 791–9.

Chua, E.F., E. Rand-Giovannetti, D.L. Schacter, M.S. Albert, and R.A. Sperling. 2004. 'Dissociating Confidence and Accuracy: Functional Magnetic Resonance Imaging Shows Origins of the Subjective Memory Experience'. *Journal of Cognitive Neuroscience*, 16, 1131–42.

Desrosières, A. 1998. *The Politics of Large Numbers: A History of Statistical Reasoning*, Harvard University Press, Cambridge.

Diaconis, P. and D.A. Freedman. 1986. 'On the Consistency of Bayes Estimates'. *Annals of Statistics*, 14, 1–26.

Diaconis, P., S. Holmes, and R. Montgomery. 2007. 'Dynamical Bias in the Coin Toss'. *SIAM Review*, 49, 211–35, doi: 0.1137/S0036144504446436.

Diaconis, P. and B. Skyrms. 2018. *Ten Great Ideas about Chance*, Princeton University Press, Princeton, NJ, and Oxford.

Frank, R.H. 2000. 'Why Is Cost-Benefit Analysis So Controversial?' *Journal of Legal Studies*, 29, 913–30.

Freedman, D.A. 1999. 'Wald Lecture: On the Bernstein-von Mises Theorem with Infinite Dimensional Parameters'. *Annals of Statistics*, 27, 1119–41.

Freedman, D.A. 2010a. 'Issues in the Foundations of Statistics: Probability and Statistical Models'. In *Statistical Models and Causal Inference: A Dialogue with the Social Sciences*, edited by D. Collier, J.S. Sekhon, and P.B. Stark, Cambridge University Press, New York, pp. 3–22.

Freedman, D.A. 2010b. 'On Types of Scientific Inquiry: The Role of Qualitative Reasoning'. In *Statistical Models and Causal Inference: A Dialogue with the Social Sciences*, edited by Collier, D., J.S. Sekhon, and P.B. Stark, Cambridge University Press, New York, pp. 337–56.

Freedman, D.A. 2010c. 'The Grand Leap'. In *Statistical Models and Causal Inference: A Dialogue with the Social Sciences*, edited by D. Collier, J.S. Sekhon, and P.B. Stark, Cambridge University Press, New York, pp. 243–54.

Freedman, D.A., and R.A. Berk. 2010. 'Statistical Models as Empirical Commitments'. In *Statistical Models and Causal Inference: A Dialogue with the Social Sciences*, edited by D. Collier, J.S. Sekhon, and P.B. Stark, Cambridge University Press, New York.

Funtowicz, S.O. and J.R. Ravetz. 1994. 'The Worth of a Songbird: Ecological Economics as a Post-Normal Science'. *Ecological Economics, 10*, 197–207.

Houser, T., S. Hsiang, R. Kopp, and K. Larsen. 2015. *Economic Risks of Climate Change: An American Prospectus*, Columbia University Press, New York.

Kahneman, D. 2011. *Thinking, Fast and Slow*, Farrar, Strauss, and Giroux, New York.

Kennedy, M.C. and A. O'Hagan. 2001. 'Bayesian Calibration of Computer Models'. *Journal of the Royal Statistical Society, B, 63*, 425–64.

Klemeš, V. 1989. 'The Improbable Probabilities of Extreme Floods and Droughts'. In *Hydrology of Disasters: Proceedings of the World Meteorological Organization*, edited by O. Starosolszky and O.M. Melder, Routledge, Abingdon, pp. 43–51.

Krug, K. 2007. 'The Relationship between Confidence and Accuracy: Current Thoughts of the Literature and a New Area of Research'. *Applied Psychology in Criminal Justice, 3*, 7–41.

Le Cam, L. 1977. 'A Note on Metastatistics, or "An Essay toward Stating a Problem in the Doctrine of Chances"'. *Synthese, 36*, 133–60.

Luce, R.D. and J.W. Tukey. 1964. 'Simultaneous Conjoint Measurement: A New Type of Fundamental Measurement'. *Journal of Mathematical Psychology, 1*, 1–27.

Mastrandrea, M.D., C.B. Field, T.F. Stocker, O. Edenhofer, K.L. Ebi, D.J. Frame, H. Held, E. Kriegler, K.J. Mach, P.R. Matschoss, G.-K. Plattner, G.W. Yohe, and F.W. Zwiers. 2010. *Guidance Note for Lead Authors of the IPCC Fifth Assessment Report on Consistent Treatment of Uncertainties: Intergovernmental Panel on Climate Change (IPCC)*. https://www.ipcc.ch/site/assets/uploads/2010/07/uncertainty-guidance-note.pdf (last accessed 27 March 2023).

Mayo, D. 2018. *Statistical Inference as Severe Testing: How to Get Beyond the Statistics Wars*, Cambridge University Press, Cambridge.

Millikan, R.A. 1913. 'On the Elementary Electrical Charge and the Avogadro Constant'. *Physical Review, 2*, 109–43.

Nature. 1978. 'Rothschild's Numerate Arrogance'. *Nature, 276*, 429, doi: 10.1038/276429a0.

O'Hagan, A. 1998. 'Eliciting Expert Beliefs in Substantial Practical Applications'. *Journal of the Royal Statistical Society. Series D (the Statistician), 47*, 21–35.

Ogasa, N. 2022. 'Mass Shootings and Gun Violence in the United States Are Increasing'. *Science News*, 26 May 2022. https://www.sciencenews.org/article/gun-violence-mass-shootings-increase-united-states-data-uvalde-buffalo (last visited 21 June 2022).

Ranson, M. 2014. 'Crime, Weather, and Climate Change'. *Journal of Environmental Economics and Management, 67*, 274–302.

Rayner, S. 2012. 'Uncomfortable Knowledge: The Social Construction of Ignorance in Science and Environmental Policy Discourses'. *Economy and Society*, *41*, 107–25, doi: 10.1080/03085147.2011.637335.

Regier, J.C. and P.B. Stark. 2015. 'Uncertainty Quantification for Emulators'. *SIAM/ASA Journal on Uncertainty Quantification*, *3*, 686–708, doi: 10.1137/130917909, http://epubs.siam.org/toc/sjuqa3/3/1.

Rhodium Group. 2014. *The American Climate Prospectus*, https://rhg.com/wp-content/uploads/2014/10/AmericanClimateProspectus_v1.2.pdf (last accessed 1 June 2022).

Saltelli, A. 2019. 'Statistical versus Mathematical Modelling: A Short Comment'. *Nature Communications*, *10*, 1–3, doi: 10.1038/s41467-019-11865-8.

Saltelli, A., G. Bammer, I. Bruno, E. Charters, M. Di Fiore, E. Didier, W.N. Espeland, J. Kay, S. Lo Piano, D. Mayo, R. Pielke Jr, T. Portaluri, T.M. Porter, A. Puy, I. Rafols, J.R. Ravetz, E. Reinert, D. Sarewitz, P.B. Stark, A. Stirling, J. van der Sluijs, and P. Vineis. 2020. 'Five Ways to Ensure that Models Serve Society: A Manifesto'. *Nature*, *582*, 482–4.

Saltelli, A., P.B. Stark, W. Becker, and P. Stano. 2015. 'Climate Models as Economic Guides: Scientific Challenge or Quixotic Quest?' *Issues in Science and Technology*, *31(3)*, Spring 2015.

Sassone, P.G. and W.A. Schaffer. 1978. *Cost-Benefit Analysis: A Handbook*, Academic Press, San Diego, CA.

Schulz, M.-A., B. Schmalbach, P. Brugger, and K. Witt. 2012. 'Analyzing Humanly Generated Random Number Sequences: A Pattern-Based Approach'. *PLoS ONE*, *7*, e41531, doi: 10.1371.

Scoones, I. and A. Stirling. 2020. *The Politics of Uncertainty*, Routledge, Abingdon; New York, doi: 10.4324/9781003023845.

Shermer, M. 2008. 'Patternicity: Finding Meaningful Patterns in Meaningless Noise'. *Scientific American*, 1 December 2008. https://www.scientificamerican.com/article/patternicity-finding-meaningful-patterns (last accessed 17 March 2023).

van der Sluijs, J.P. 2016. 'Numbers Running Wild'. In *Science on the Verge*, edited by A. Benessia, S. Funtowicz, M. Giampietro, A. Saltelli, Â.G. Pereira, J.R. Ravetz, R. Strand, and J.P. van der Sluijs, Consortium for Science, Policy, and Outcomes, Tempe, AZ, and Washington, DC, pp. 151–88.

Stark, P.B. 2015. 'Constraints versus Priors'. *SIAM/ASA Journal of Uncertainty Quantification*, *3*, 586–98, doi: 10.1137/130920721.

Stark, P.B. 2022 'Reproducibility, P-Values, and Type III errors: response to Mayo 2022'. *Conservation Biology*, *36*(5), doi: 10.1111/cobi.13986.

Stark, P.B. and D.A. Freedman. 2010. 'What Is the Chance of an Earthquake?' In *Statistical Models and Causal Inference: A Dialogue with the Social Sciences*, edited by D. Collier, J.S. Sekhon, and P.B. Stark, Cambridge University Press, New York, pp. 115–30.

Stark, P.B. and R. Freishtat. 2014. 'An Evaluation of Course Evaluations'. *ScienceOpen Research*, doi: 10.14293/S2199-1006.1.SOR-EDU.AOFRQA.v1.

Stark, P.B. and A. Saltelli. 2018. 'Cargo-Cult Statistics and Scientific Crisis'. *Significance*, *15*(4), 40–3, doi: 10.1111/j.1740-9713.2018.01174.x. Fully referenced version: https://www.significancemagazine.com/593 (last accessed 30 May 2022).

Stark, P.B. and L. Tenorio. 2010. 'A Primer of Frequentist and Bayesian Inference in Inverse Problems'. In *Large-Scale Inverse Problems and Quantification of Uncertainty*, edited by L. Biegler, G. Biros, O. Ghattas, M. Heinkenschloss, D. Keyes,

B. Mallick, L. Tenorio, B. van Bloemen Waanders and K. Willcox, John Wiley and Sons, New York, pp. 9–32.

Stein, S. and J.L. Stein. 2013. 'Shallow versus Deep Uncertainties in Natural Hazard Assessments'. *Eos, 94(4)*, 133–40.

Stigler, S.M. 1986. *The History of Statistics: The Measurement of Uncertainty before 1900*, Harvard University Press, Cambridge, MA.

Taleb, N.N. 2007. *The Black Swan: The Impact of the Highly Improbable*, Random House, New York.

Tversky, A. and D. Kahneman. 1975. 'Judgment under Uncertainty: Heuristics and Biases'. *Science, 185*, 1124–31.

Watson, R.T., P.S. Kolar, M. Ferrer, T. Nygård, N. Johnston, W.G. Hunt, H.A. Smit-Robinson, C.J. Farmer, M. Huso, and T.W. Katzner. 2018. 'Raptor Interactions with Wind Energy: Case Studies from Around the World'. *Journal of Raptor Research, 52(1)*, 1–18, doi: 0.3356/JRR-16-100.1.

PART II

THE RULES

3

Mind the framings

Match purpose and context

*Monica Di Fiore, Marta Kuc-Czarnecka, Samuele Lo Piano, Arnald Puy,
and Andrea Saltelli*

Introduction

In several ways, the following chapters will allude to the concept of frames.

The assumptions of a mathematical model tell us what it can explain and what it cannot (see chapter 5). The trade-off between the usefulness of a mathematical model and the scale of complexity it seeks to capture further conditions the narrative, highlighting or downsizing cognitive elements relevant to responsible modelling (see 'Mathematics and tales' in chapter 4). Yet, what determines the choice of a model, its assumptions, and its level of complexity? Answering these questions requires a more general perspective that involves frames—that is, the complex of cognitive schemes through which we reason and which we use to make sense of the world.

The notion of the frame owes much to the tradition of social studies. Several disciplines have underlined the importance of worldviews and their causal relationships in giving us the coordinates to classify the situations in which we find ourselves. Since public policies are constructs of considerable complexity, *policy inquiry* has paid great attention to the frames we use to organize information, ideas, and beliefs. For example, Campbell (2002), analysing the effects of ideas, worldviews, and cognitive paradigms on policy-making, defines frames as 'normative and sometimes cognitive ideas that are located in the foreground of policy debates'.

In a perceptive work devoted to frame-critical policy analysis and frame-reflective policy practice, Rein and Schön (1996) recognize several different ways of looking at frames, as 'distinct but mutually compatible images, rather than competing conceptions'. These images are 'a scaffolding (an inner structure), a boundary that sets off phenomena from their contexts (like picture frames), a cognitive/appreciative schema of interpretation [...] or a generic diagnostic/prescriptive story'. According to the authors, all these images 'rest on a

Monica Di Fiore et al., *Mind the framings*. In: *The politics of modelling*. Edited by: Andrea Saltelli and Monica Di Fiore, Oxford University Press. © Oxford University Press (2023). DOI: 10.1093/oso/9780198872412.003.0003

common insight: there is a less visible foundation—an "assumptional basis"—that lies beneath the more visible surface of language or behaviour, determining its boundaries and giving it coherence'.

When discussing the frames that may uphold a given modelling exercise, one problem is the abundance of lenses that one can bring to bear on the issue. For example, sociotechnical imaginaries—that is, shared visions and expectations of how scientific and technological progress will shape human futures (Jasanoff and Kim 2015), refer to underlying frames about public goals and common futures. Here adopting a utopian or dystopian vision of the impact of a new technology can clearly determine the narrative upheld by a modelling exercise. A historical case where scholars targeted scenarios linked to the introduction of genetically modified substances is the Public Perceptions of Agricultural Biotechnologies (PABE) study, where an original techno-optimistic framing focusing on the concept of risk was reversed by calling attention to concepts of public agency and democratic accountability (Marris 2001).

Also worth mentioning is the theory of 'orders of worth' proposed by the French school of sociology (Boltanski and Thévenot 2006). According to this theory, different actors and institutions support and aliment their legitimacy by engaging in discourses that suit their objectives. Orders of worth represent higher principles that correspond to an existing social order and on which one can act as to justify a given premise. Worth exists as a plurality, for example one can appeal to a civic worth (collective solidarity and social justice), a green worth (environmental concern), or a domestic worth (traditional values), and so on (Thévenot 2022). One or another of these 'worths' can be appealed to and used as a ground for negotiation or contestation among different stakeholders. Different orders of worth correspond to the need to quantify or model different objects. Making these orders explicit and visible is among the tasks suggested in sensitivity auditing (see 'Introduction' in chapter 8).

In the end, whether a model can be judged 'responsible' or 'fair' by its intended audience, and by society, largely depends upon its frame. To access a frame, one needs to overcome the supposed neutrality of the model, and to be perceptive of a full spectrum of possible motivations of its developers, including rhetorical or instrumental uses. Frames condition both research agendas and policy implications (Rein and Schön 1996).

Neglect of the frames underlying a given modelling exercise may lead to what several scholars call *hypocognition*. For Saltelli and Giampietro (2017), one cause of hypocognition is 'a lack of the appropriate frames that would be needed to become aware of a problem. [...] Lacking these frames we simply do not see it.'

The mathematical problem formulation

In 1980, Majone and Quade (1980) published a volume devoted to 'pitfalls of analysis'. The contribution addressed system analysis in support of policy, paying considerable attention to the issue of predictive modelling. In chapter 3 of *Pitfalls in Analysis*, Quade writes:

> Problem formulation, involving as it does the simplification of a subset of the real world selected for study, may, in fact, be considered part of the modeling process because the assumptions about the problem setting affect the predictions of the consequences, with the model being the analyst's conception of the factors and relationships significant to his investigation.
>
> (Quade 1980)

The quote from Quade stresses the close relationship between the context and the purpose of a model: in other words, the scope for its formulation and the activity of modelling proper. The quote also highlights the linkages between the apparently technical activity of developing a model and the context in which this activity occurs (Box 3.1), as in the *Manifesto*'s 'Mind the Framing' rule (Saltelli et al. 2020a). Results from models will at least partly reflect the interests, disciplinary orientations, and biases of the developers, as well as the demands put on the developers by those who originate or own the problem. Therefore, pitfalls in modelling may result from a poor match between the demand and the supply side of a modelling exercise. Other pitfalls originate when, at the other extreme, the models' developers adopt or internalize the demand put on them to the point that they lose a critical view of the problem. In this latter case, the frame disappears into the background, and the model, clothed in an apparent neutrality, presents itself as the logical and rational choice one needs to take to tackle the problem at hand.

Quade (1980) highlighted other aspects that inspired the *Manifesto*, such as the ambition to capture reality with increasingly complex models, or the desire to substitute the model for the analysis, forgetting that '[j]ust as analysis is the servant of judgment, computers are a servant of analysis' (Enthoven and Smith 2005: 63; Box 3.1).

Box 3.1 Pitfalls in formulation and modelling

Pitfalls in formulation

Insufficient attention to formulation
Unquestioning acceptance of stated goals and constraints
Measuring achievement by proxy
Misjudging the difficulties
Bias

Pitfalls in modelling

Equating modelling with analysis
Improper treatment of uncertainties
Attempting to really simulate reality
Belief that a model can be proved correct
Neglecting the by-products of modelling
Overambition
Seeking academic rather than policy goals
Internalizing the policy maker
Not keeping the model relevant
Not keeping the model simple
Capture of the user by the modeller

Source: (Quade 1980)

Since the ambition of the present work is to contribute to a reciprocal under-standing between models and society, we offer hereafter some reflections that may help the reader to 'read' through the flesh of the model into the bones of its frame. This is a guide not only for the users of models, but for modellers as well.

The model of everything: Spoken and unspoken frames

No one model can serve all purposes. Diligent practitioners know that when a model built for one purpose is then used for another, it needs to be reassessed against the new task, and in some instances even rebuilt, given the change of circumstances (Edmonds et al. 2019). This is not the standard practice in math-ematical modelling and, as noted by Quade, both modellers and their customers may end up being captured by a model that establishes itself as the workhorse for all jobs.

Discussing frames, it comes naturally to ponder a master narrative—perhaps a meta-frame—often found among modellers: that of the perfect model, one that will take time and effort to construct, but that once built will answer all questions relative to a city, a hydrological basin, or even the planet. Literature has mused on the concept since Douglas Adams's (1979) 'Deep Thought'—a mega computer among the characters of his novel, *The Hitchhiker's Guide to the Galaxy*—announced that an even bigger machine would be built, so large that 'it would resemble a planet, and be of such complexity that organic life itself would become part of its operating matrix'.

It is a fact that there are models so precise and accurate as to inspire awe. As noted by John Kay and Mervyn King in their book on radical uncertainty, there are cases when models achieve extraordinary feats. NASA modellers could position the probe MESSENGER around Mercury after five billion miles and six and a half years of travel (Kay and King 2020). The trajectory of MESSENGER included several revolutions around the Sun and flyby manoeuvres near Earth, Venus, and Mercury to reduce the speed relative to Mercury, before final insertion into the planet's orbit. These feats are possible due to knowledge of the law of motion and the related parameters, the stationarity of the phenomenon, and the lack of human influence in the targeted system. The example is extraordinary in that these conditions are rarely met in modelling.

Model-knowing, in fact, is conditional. It is based on knowing a certain number of relevant properties of the system modelled that are needed for the construction of the model itself. Lacking this knowledge results in what John Kay (2012) calls 'bogus models'—that is, models whose relevant inputs are simply made up by the analysts. For Kay, three instances or classes of models that fit into this category are: (i) *WebTag*, a framework used in the UK for appraising transport projects; (ii) public sector comparator models, which are used to assess potential private financial initiative projects; and (iii) value at risk modelling, which is used for risk management in banks.[1] Note that all these models appear at least nominally relevant for the taking of important policy or financial decisions.

For Amartya Sen (1990) and Robert Salais (2022), the quality of a quantification that assists in taking political actions—of which predictive mathematical modelling is an instance—needs to be justified on both technical and normative grounds. In relation to the frames and assumptions underlying a modelling exercise, quality is violated when an input number is made up in order to run a model, as in Kay's (2012) example of the average number of passengers sitting in a car decades into the future. When instead the model makes assumptions that are only relevant and useful to some and irrelevant to others, it is the normative dimension of quality that is violated. This is especially the case when a model fails to identify winners and losers of a given transaction (Jasanoff 2007).

Consider GENESIS, a model used to simulate the behaviour of shorelines and their response to beach replenishment or other coastal engineering interventions (Young et al. 1995; see also Box 1.1). The model is used by the US Army Corps

---

[1] The reader is referred to the work of John Kay (2012) for a discussion of these model-related references.

of Engineers, which adopts an engineering-framed analytics. This is an example of a discipline-inspired frame, where the pitfalls of spoken and unspoken frames become evident.

This model applies an engineering mindset to an ecological problem. Orrin H. Pilkey (2000: 163) considers this one source of the problem in GENESIS when he notes that 'it is difficult to transfer the mathematical modelling approach from predicting the behavior of steel and concrete structures to predicting the course of earth surface processes'. By the way it is designed, GENESIS appears particularly flexible in its operations, so as to confirm 'the results that the coastal "expert" employing its services expected—a way of backing up one's judgment with what appear to be real numbers' (Pilkey 2000: 172).

This is not the only vulnerability of this model, as the existing critiques show (Young et al. 1995, Pilkey and Pilkey-Jarvis 2009). GENESIS uses averages to parameterize natural phenomena, such as storms. The pitfall of this approach is that beaches are not eroded by average storms, but by exceptional ones. One can find parallels to this predicament when using averages in the analysis of financial markets, where for Nassim Taleb (2011) the most significant changes occur in the course of extraordinary events. In other words, models may be built to reproduce what happens when little happens. As noted by May (2004) in relation to models used in biology, one may find 'excruciating abundance of detail in some aspects, whilst other important facets of the problem are misty or a vital parameter is uncertain to within, at best, an order of magnitude'.

The epistemic status of a mathematical model: Between rhetorical and technical neutrality

When compared to other families of quantification—that is, statistical indicators, rating and ranking (discussed in chapter 9), statistical analysis (discussed in chapter 2), and algorithmic developments, modelling is perhaps the one that enjoys the highest epistemic status. Models have easier access to claims of neutrality than other instances of quantification—what Porter (1994) calls the rhetoric of impersonality comes easily to the product of mathematical abstraction. Such mathematical accuracy corresponds well to the 'Cartesian dream' (Davies and Hersh 1986), the idea that there is an underlying mathematical structure governing the empirical reality that is prone to be deciphered with scientific methods. Yet the epistemological and practical grounds of this dream tend to crack when models apply to complex, non-deterministic processes influenced by humans and/or characterized by uncertainties and indeterminacies. Models enjoy licences, such as creating 'probabilities' from 'observed rates', which would be frowned upon by statisticians, or performing experiments *in silico* that are presented as evidence of equivalent experiments in the real world (see 'Signifying nothing' in the Preface).

Models have become so adept at 'navigating the political' (van Beek et al. 2022) that we hardly notice anymore when a political problem has been transformed by a model into a technical one. As an example, food security can be presented as a problem of technology (for example better crops or irrigation methods), obscuring issues of geopolitics and power unbalance (see 'Food security' in chapter 8).

In sensitivity auditing (Saltelli et al. 2013), these two dimensions of quality are covered by different rules: the first, named 'Assumption Hunting', invites the modellers and its customers to retrace all that was assumed in or assumed out, to test for relevant omissions or arbitrariness that invalidate the inference (see chapter 5). The second rule, named 'Do the Right Sums', alludes to the right sums being used to tackle the wrong problem. Doing the right sums has to do with refraining from purported neutrality (Saltelli et al. 2020b), opening the space of problem solutions to a plurality of stakeholders, and paying attention to what is simply unfeasible or unviable, abandoning an analyst's obsession with finding the optimal solution to a problem.

The analytical tool used to investigate a practical problem largely determines the nature of the conclusions. The German sociologist Ulrich Beck (1992) made the point in his 1986 work *Risk Society*:

> It is not uncommon for political programs to be decided in advance simply by the choice of what expert representatives are included in the circle of advisers.

The issue of non-neutrality of technique, and of how methodological choices condition the generation of narratives, is a recurring theme in both ecological and political sciences. One solution is to look at relevant and/or conflicted issues by approaching the problem with lenses coming from different disciplines or intellectual traditions. This also includes those that are less often seen in action, ranging from non-Ricardian economics to feminist economics, from bioeconomics to post-normal science (Saltelli et al. 2020b, Saltelli et al. 2022a). The non-neutrality of technique is also relevant due to the knowledge asymmetry between producers and users of mathematical modelling. Modellers know vastly more about the conditionality of the predictions of their models than is usually communicated to the users/customers of the analysis (Jakeman, Letcher, and Norton 2006). This is not necessarily the result of bad faith but a consequence of the several hidden, non-explicit assumptions a model is based on.

An example: The value of a statistical life

In the early phases of the COVID-19 pandemic, it was posited that social distancing in the US would lead to a net benefit of $5.2 trillion (Thunstrom et al. 2020). This result was obtained using the concept of value of a statistical life (VSL). VSL is used in several applications, including to decide upon compensatory damages in actuarial sciences (Viscusi 2008). Even in this field, the framing offered by VSL, which allows a neat arithmetic to settle complicated cases, has been met with controversy. When an airplane falls, the families of the victims may be given different compensations depending on the nationality of the deceased or the jurisdiction in which the case is filed (Linshi 2015). To add to the confusion, even in a single country, the US, different regulatory agencies adopt different VSLs (Viscusi 2008).

These situations are well captured by Power's (2004) remark that 'Political power can be understood as the ability to make even controversial counting and measurement systems appear natural and unavoidable, preventing the widespread institutionalization of distrust in numbers and supporting a variety of schemes for monitoring and control'. This author describes in detail the dialectic inherent to quantification: it exists as a tool to both achieve democratic accountability and exercise a power that appears impersonal. Quantified analyses may thus appear as 'fatal remedies' that create incentives to undermine the very activity being measured, thus opening the door to all forms of management and gaming (see 'Modelling and mutually reinforcing authoritarianisms' in chapter 7).

Theodore Porter (1994) describes the concept of VSL as a 'matter of continuing controversy', starting with an incursion into satire in the seventeenth century, when Johnathan Swift poked fun at the influential William Petty for his idea that the Irish population—save for a few cowherds—should be forcibly moved to England since their higher value there compensated for the cost of the shipping (Porter 1994). Under the powerful driver of standardization, VSL has remained a fundamental ingredient of cost–benefit analysis, where it was used for most of the twentieth century, for example for road or water projects, taking as input the numbers of the insurance sector. These numbers mostly reflected the discounted value of lost earnings, with the consequent normative problems of valuing breadwinners more than their spouses or the elderly.

It is clear that VSL and the associated cost–benefit analyses are economic solutions to political problems (Porter 1994). These approaches can be loved or hated depending upon the use to which they are put, such as improving efficiency, increasing profit, promoting hygiene and sanitation, achieving better education, or upholding the care of the environment. Like many other instances of quantification, quantitative assessments and standards may be introduced so as to fight knowledge asymmetry and arbitration from the most powerful, as during the French Revolution's development of standardized measure of a bushel to prevent the abuse of farmers by landlords (see 'The numbers critique' in chapter 6). Over time, the same measures go on to be contested, when their use comes to be identified with a system that is perceived as invasive or oppressive (Porter 1994):

> It is, on the whole, external pressure that has led to the increasing importance of calculation in administration and politics. Those whose authority is suspect,

and who are obliged to deal with an involved and suspicious public, are much more likely to make their decisions by the numbers than are those who govern by divine or hereditary right. It is not by accident that the authority of numbers is linked to a particular form of government: representative democracy.

It is not surprising that all that is technical in a cost–benefit analysis can become the subject of conflict and manipulations. For example, those who favour the implementation of a project want discount rates to be low, so that the benefits will keep accruing over time, and those who oppose the same project want these to be high (Porter 1994).

These conflicted and context dependent uses of models suggest that users should exercise circumspection, while developers should come prepared for opposition. Both should weigh the benefits of standardization against the cost of its absence. This is needed even more, since analyses such as those just described are unlikely to run out of fashion. Stakeholders' participation, with all its difficulties and limits, appears to be an important tool to monitor use and prevent abuse.

Involving stakeholders: A new social contract for models

Model validation and verification approaches can be divided into the purely technical (for example in using sensitivity analysis or sharing the code in repositories) and those which we could call reflexive (Eker et al. 2018).

Reflexive approaches include looking at a richer set of attributes of the modelling process, such as framings, expectations and motivations of the developers, tacit assumptions, implicit biases in the use of a particular modelling approach, care in not overinterpreting the results, and so on.

Examples of stakeholder involvement in the modelling process can be found in basin management (Lane et al. 2011), fisheries (Röckmann et al. 2012), medicine (Miedema 2016), and mathematical modelling in general (Refsgaard et al. 2006). This has led to the coining of the term 'participatory modelling'. For instance, Lane et al. (2011) discuss a case where models were co-produced by the experts with the involved local community in a way that led to unexpected results. The problem tackled was how to deal with recurring flooding in the small market town of Ryedale, North Yorkshire, UK.

According to the authors, the study revealed 'a deep and distributed understanding of flood hydrology across all experts, certified and uncertified'. In other words, knowledge was not all on the side of the experts. Despite the expertise of modellers and hydrologists involved in the task, an alternative intervention—upstream storage of floodwaters—was only 'discovered' when local stakeholders were brought into the modelling process. The model ignored this option because of the 'use of a pit-filling algorithm that made sure that all water flows downhill'. The experts had to construct a brand-new model to account for upstream water storage processes following the suggestion of the local community. In cases like this, sources of tacit or forgotten knowledge both among the experts and the lay public can only be discovered when heads are banged together.

Another example is that of the project JAKFISH (Judgement and Knowledge in Fisheries Involving Stakeholders), which by design was about participatory modelling (Röckmann et al. 2012). The project looked at the potential risks for stocks and fisheries associated with biological, socioeconomic, and management aspects.

The study adopted a stance inspired by post-normal science (see 'The epistemic background of sensitivity auditing' in chapter 8), recognizing the existence of pre-conceived ideas among both experts and their constituency: 'Stakeholders have an agenda, and at the same time, scientists have scientific agendas or at least personal scientific ambitions' (Röckmann et al. 2012). In these situations, agreeing on the nature of the problem and on the desired 'end in sight' are important preliminary steps in the analysis. Also in this case, models had to be built on an ad hoc basis during the negotiation, keeping in mind the trade-off between model complexity and model transparency (see chapter 4).

Proper consideration of the uncertainty also led to different results: 'lower fishing mortality targets were required to maintain pre-agreed stock levels with a certain probability than if no uncertainty was considered' (Röckmann et al. 2012).

Although it is time- and resource-intensive, participatory modelling may allow the testing of different models and solutions, as discussed in another fisheries example characterized by a careful appreciation of uncertainties (Fulton et al. 2014). This study led the Australian Fisheries Management Authority to reconsider and eventually rewrite the fisheries management plan for the Australian Southern and Eastern Scalefish and Shark Fishery.

Conclusion

As discussed in this chapter, models are powerful tools for the exercise of the rhetoric of objectivity, holding the promise of the 'view from nowhere'—to use the expression of philosopher Thomas Nagel (1989)—which elevates a policy decision above the fray of contrasting interests and stakes. Only a clear perception of the underlying frames may separate a model from the dangers of such a rhetorical use.

When discussing frames, we have noted how stakeholders involvement can lead to the creation of shared meanings. The example of the value of life, 'perhaps the best example of a calculation which respects nobody's meanings' (Porter 1994), in a sense implies the need for meaning to be created through exchange among different constituencies.

It needs to be stressed that participatory modelling, as all forms of participation, is far from easy and comes with costs and risks. At its core, participation implies the existence of publics that are capable and willing to engage. In the social and political sciences, this has been a contentious point ever since the debates between the pragmatist philosopher John Dewey and political analyst Walter Lippmann: can the public express agency? At what risk? The concerned reader may refer to Philip Mirowski's 2020 essay, in which he revisits the history of these issues, as well as Saltelli et al. (2022b), which investigates their possible relation to the phenomenon of regulatory capture, whereby the public itself may become hostage to private interests.

While participation is inherently risky, its absence can easily lead to worse outcomes. For example, Sámi reindeer herders' situation can only be fully appreciated by engagement with those who are directly interested. For Tyler et al. (2007), loss of habitat, economic predation, and legal frameworks 'potentially dwarf the putative effects of projected climate change on reindeer pastoralism'. In addition:

> The validity and legitimacy of reducing a complicated system to something simple and, therefore, amenable to assessment was wholly dependent on the participation at the outset of herders themselves. It is they, rather than outsiders, who can best decide what factors, or what suites of factors, influence reindeer pastoralism: nobody, save herders themselves, can legitimately make the selection. Despite its orthodox format, therefore, the resulting conceptual model, developed through an interdisciplinary and intercultural effort [...] represented an integration of empirical data and herders' knowledge.

The same authors note how the herders embody generations of experience with coping with a changing planet and socioeconomic conditions: 'Herders integrate bodies of knowledge gathered over time spans that far exceed significant periods of climate change. It would not be possible, using the traditional methods of the natural sciences, to gather comparable bodies of knowledge by direct observation at less than exorbitant cost.'

References

Adams, D. 1979. *The Hitchhiker's Guide to the Galaxy*, Pan Books, London.

Beck, P.U. 1992. *Risk Society: Towards a New Modernity*, Sage Publications, Newbury Park, CA (Published in association with Theory, Culture & Society).

van Beek, L., J. Oomen, M. Hajer, P. Pelzer, and D. van Vuuren. 2022. 'Navigating the Political: An Analysis of Political Calibration of Integrated Assessment Modelling in Light of the 1.5 °C Goal'. *Environmental Science & Policy*, 133, 193–202, doi: 10.1016/j.envsci.2022.03.024.

Boltanski, L. and L. Thévenot. 2006. *On Justification: Economies of Worth*, 1st edition, translated by C. Porter, Princeton University Press, Princeton, NJ.

Campbell, J.L. 2002. 'Ideas, Politics, and Public Policy'. *Annual Review of Sociology*, 28(1), 21–38, doi: 10.1146/annurev.soc.28.110601.141111.

Davis, P.J. and R. Hersh. 1986. *Descartes' Dream: The World According to Mathematics*, Penguin Books, London.

Edmonds, B., C. Le Page, M. Bithell, E. Chattoe-Brown, V. Grimm, R. Meyer, C. Montañola-Sales, P. Ormerod, H. Root, and F. Squazzoni. 2019. 'Different Modelling Purposes'. *Journal of Artificial Societies and Social Simulation*, 22(3), 6, doi: 10.18564/jasss.3993.

Eker, S., E. Rovenskaya, M. Obersteiner, and S. Langan. 2018. 'Practice and Perspectives in the Validation of Resource Management Models'. *Nature Communications*, 9(1), 5359, doi: 10.1038/s41467-018-07811-9.

Enthoven, A.C. and K.W. Smith. 2005. *How Much Is Enough? Shaping the Defense Program, 1961–1969*, Harper and Row, New York.

Fulton, E.A., A.D.M. Smith, D.C. Smith, and P. Johnson. 2014. 'An Integrated Approach Is Needed for Ecosystem Based Fisheries Management: Insights from Ecosystem-Level Management Strategy Evaluation'. *PLoS ONE*, 9(1), e84242, doi: 10.1371/journal.pone.0084242.

Jakeman, A.J., R.A. Letcher, and J.P. Norton. 2006. 'Ten Iterative Steps in Development and Evaluation of Environmental Models'. *Environmental Modelling & Software*, 21(5), 602–14.

Jasanoff, S. 2007. 'Technologies of Humility'. *Nature*, 450(7166), 33, doi: 10.1038/450033a.

Jasanoff, S. and Kim, S.-H. 2015. Dreamscapes of Modernity: Sociotechnical Imaginaries and the Fabrication of Power, The University of Chicago Press, Chicago, IL.

Kay, J. 2012. 'Bogus Modelling Discredits Evidence-Based Policy'. *The London School of Economics* [Blog]. http://blogs.lse.ac.uk/politicsandpolicy/bogus-modelling-kay/ (last accessed 22 March 2023).

Kay, J.A. and M.A. King. 2020. *Radical Uncertainty: Decision-Making beyond the Numbers*, W. W. Norton & Company, New York.

Lane, S.N., N. Odoni, C. Landström, S.J. Whatmore, N. Ward, and S. Bradley. 2011. 'Doing Flood Risk Science Differently: An Experiment in Radical Scientific Method'. *Transactions of the Institute of British Geographers*, 36(1), 15–36. doi: 10.1111/j.1475-5661.2010.00410.x.

Linshi, J. 2015. 'Germanwings Plane Crash: How Much Compensation for Victims' Families? | Time'. *Time*, 31 March 2015. https://time.com/3763541/germanwings-plane-crash-settlement.

Majone, G. and E.S. Quade. 1980. *Pitfalls of Analysis*. International Institute for Applied Systems Analysis. Chichester, UK: Wiley, Chichester, UK.

Marris, C. 2001. *Final Report of the PABE Research Project Funded by the Commission of European Communities, Contract number: FAIR CT98-3844 (DG 12 - SSMI)*.

May, R.M. 2004. 'Uses and Abuses of Mathematics in Biology', *Science*, 303(5659), 790–3. doi: 10.1126/science.1094442.

Miedema, F. 2016. 'To Confront 21st Century Challenges, Science Must Rethink Its Reward System'. *The Guardian*, 12 May 2016. https://www.theguardian.com/science/political-science/2016/may/12/to-confront-21st-century-challenges-science-needs-to-rethink-its-reward-system (last accessed 22 March 2023).

Mirowski, P. 2020. 'Democracy, Expertise and the Post-Truth Era: An Inquiry into the Contemporary Politics of STS'. *Academia.eu*, April. https://www.academia.edu/42682483/Democracy_Expertise_and_the_Post_Truth_Era_An_Inquiry_into_the_Contemporary_Politics_of_STS (last accessed 22 March 2023).

Nagel, T. 1989. *The View from Nowhere*, revised edition, Oxford University Press, New York; London.

Pilkey, O.H. 2000. 'What You Know Can Hurt You: Predicting the Behavior of Nourished Beaches'. In *Prediction: Science, Decision Making, and the Future of Nature*, edited by D. Sarewitz, R. A. Pielke Jr, and R. Byerly, Island Press, Washington, DC, 159–84.

Pilkey, O.H. and L. Pilkey-Jarvis. 2009. *Useless Arithmetic: Why Environmental Scientists Can't Predict the Future*, Columbia University Press, Washington, DC.

Porter, T.M. 1994. 'Objectivity as Standardization: The Rhetoric of Impersonality in Measurement, Statistics, and Cost-Benefit Analysis'. In *Rethinking Objectivity*, edited by A. Megill, Duke University Press, Durham, NC, pp. 197–237.

Power, M. 2004. 'Counting, Control and Calculation: Reflections on Measuring and Management'. *Human Relations*, 57(6), 765–83, doi: 10.1177/0018726704044955.

Quade, E.S. 1980. 'Pitfalls in Formulation and Modeling'. In *Pitfalls of Analysis*, edited by G. Majone and E.S. Quade, International Institute for Analysis, Wiley, Chichester UK, pp. 23–43.

Refsgaard, J.C., J.P. van der Sluijs, J. Brown, and P. van der Keur. 2006. 'A Framework for Dealing with Uncertainty Due to Model Structure Error'. *Advances in Water Resources*, 29(11), 1586–97.

Rein, M. and D. Schön. 1996. 'Frame-Critical Policy Analysis and Frame-Reflective Policy Practice'. *Knowledge and Policy*, 9(1), 85–104. doi: 10.1007/BF02832235.

Röckmann, C., C. Ulrich, M. Dreyer, E. Bell, E. Borodzicz, P. Haapasaari, K.H. Hauge, D. Howell, S. Mäntyniemi, D. Miller, G. Tserpes, and M. Pastoors. 2012. 'The Added Value of Participatory Modelling in Fisheries Management: What Has Been Learnt?' *Marine Policy*, 36(5), 1072–85. doi: 10.1016/j.marpol.2012.02.027.

Salais, R. 2022. '"La donnée n'est pas un donné": Statistics, Quantification and Democratic Choice'. In *The New Politics of Numbers: Utopia, Evidence and Democracy*, edited by A. Mennicken and R. Salais, Palgrave Macmillan, London, pp. 379–415.

Saltelli, A., Â. Guimarães Pereira, J P. van der Sluijs, and S. Funtowicz. 2013. 'What Do I Make of Your Latinorum? Sensitivity Auditing of Mathematical Modelling'. *International Journal of Foresight and Innovation Policy*, 9(2/3/4), 213–34, doi: 10.1504/IJFIP.2013.058610.

Saltelli, A., G. Bammer, I. Bruno, E, Charters, M. Di Fiore, E. Didier, W. Nelson Espeland, J. Kay, S. Lo Piano, D. Mayo, R. Pielke Jr, T.M. Portaluri, T. Porter, A. Puy, I. Rafols, J.R. Ravetz, E. Reinert, D. Sarewitz, P.B. Stark, A. Stirling, J. van der Sluijs, and P. Vineis. 2020a. 'Five Ways to Ensure that Models Serve Society: A Manifesto'. *Nature*, 582, 482–4, doi: 10.1038/d41586-020-01812-9.

Saltelli, A., L. Benini, S. Funtowicz, M. Giampietro, M. Kaiser, E. Reinert, and J P. van der Sluijs. 2020b. 'The Technique Is Never Neutral: How Methodological Choices Condition the Generation of Narratives for Sustainability'. *Environmental Science and Policy*, 106, 87–98.

Saltelli, A., M. Kuc-Czarnecka, S. Lo Piano, M.J. Lőrincz, M. Olczyk, A. Puy, E. Reinert, S.T. Smith, and J.P. van der Sluijs. 2023. 'Impact Assessment Culture in the European Union. Time for Something New?' *Environmental Science & Policy*, 142, 99–111, doi: 10.1016/j.envsci.2023.02.005.

Saltelli, A., D. Dankel, M. Di Fiore, N. Holland, and M. Pigeon. 2022b. 'Science, the Endless Frontier of Regulatory Capture'. *Futures*, 135, 102860. doi: 10.1016/j.futures.2021.102860.

Saltelli, A. and M. Giampietro. 2017. 'What Is Wrong with Evidence Based Policy, and How Can It Be Improved?' *Futures*, *91*, 62–71.

Sen, A. 1990. 'Justice: Means versus Freedoms', *Philosophy & Public Affairs*, *19*(2), 111–21.

Taleb, N.N. 2011. *The Black Swan: The Impact of the Highly Improbable*. Allen Lane, London.

Thévenot, L. 2022. 'A New Calculable Global World in the Making: Governing Through Transnational Certification Standards'. In *The New Politics of Numbers*, edited by A. Mennicken and R. Salais, Palgrave Macmillan, London, pp. 197–252.

Thunstrom, L., S.C. Newbold, D. Finnoff, M. Ashworth, and J.F. Shogren. 2020. 'The Benefits and Costs of Using Social Distancing to Flatten the Curve for COVID-19'. *Journal of Benefit-Cost Analysis*, *11*(2), 179–95, doi: 10.1017/bca.2020.12.

Tyler, N.J.C., J.M. Turi, M.A. Sundset, K. Strøm Bull, M.N. Sara, E. Reinert, N. Oskal, C. Nellemann, J.J. McCarthy, S.D. Mathiesen, M.L. Martello, O.H. Magga, G.K. Hovelsrud, I. Hanssen-Bauer, N.I. Eira, I.M.G. Eira, and R.W. Corell. 2007. 'Saami Reindeer Pastoralism under Climate Change: Applying a Generalized Framework for Vulnerability Studies to a Sub-Arctic Social–Ecological System'. *Global Environmental Change*, *17*(2), 191–206, doi: 10.1016/j.gloenvcha.2006.06.001.

Viscusi, W.K. 2008. 'The Flawed Hedonic Damages Measure of Compensation for Wrongful Death and Personal Injury'. *Journal of Forensic Economics*, *20*(2), 113–35, doi: 10.5085/0898-5510-20.2.113.

Young, R.S., O.H. Pilkey, D.M. Bush, and E.M. Thieler. 1995. 'A Discussion of the Generalized Model for Simulating Shoreline Change (GENESIS)'. *Journal of Coastal Research*, *11*(3), 875–86.

4

Mind the hubris

Complexity can misfire

Arnald Puy and Andrea Saltelli

Mathematics and tales

Possibly the most famous quote of the French philosopher Descartes (2006: 51) is that of man as 'master and possessor of nature', an individual that uses the power of mathematical reason to decode the natural order and improve the human condition. Technically, Descartes saw more clearly than others how applying algebra to solve geometrical problems would open the door to the solution of practical challenges. This vision came to Descartes in a dream and has fuelled the extraordinary development that science and technology has experienced over the last three centuries. In fact, Descartes' dream has been so successful that 'it would not be a mistake to call our age and all its scientific aspirations Cartesian' (Davis and Hersh 1986: 260).

Descartes' vision of a mechanical world dominated by mathematical laws prone to being deciphered through scientific inquiry was also shared by philosophers such as Galilei, Leibniz, Laplace, and Condorcet (Funtowicz and Pereira 2015). Underlying this premise is the platonic idea of a mathematical truth existing 'out there' that provides structure to the universe independently of the individual and their cognitive abilities. Some significant defenders of this view in the nineteenth and twentieth centuries were Frege (2007) and Gödel (1995), who argued against mathematics being a product of the human psyche and endorsed the Platonistic view as the 'only one tenable'. Years later, Benacerraf inspired one of the most famous epistemological objections to mathematical Platonism by arguing that Platonists cannot explain mathematical beliefs because mathematical objects are abstract and hence not causally active—nothing connects the knowledge holder with the object known (Benacerraf 1973, Field 1989). For Balaguer (1998: 176), the debate between mathematical Platonists and anti-Platonists may not be solvable, given that both sides hold convincing arguments and are perfectly workable philosophies of mathematics.

Arnald Puy and Andrea Saltelli, *Mind the hubris*. In: *The politics of modelling*. Edited by: Andrea Saltelli and Monica Di Fiore, Oxford University Press. © Oxford University Press (2023). DOI: 10.1093/oso/9780198872412.003.0004

And yet the assumption that mathematical entities objectively exist for us to appraise through scientific inquiry seems to have sidelined the alternative in the field of mathematical modelling. This is apparent in the trend towards ever more complex models characterizing natural science fields such as hydrology or climate studies: models get bigger to accommodate the newly acquired knowledge on previously hidden mechanisms that had been brought to light by scientific methods and state-of-the-art technologies.[1] Uncertainties are regarded as mostly epistemic and prone to being overcome or significantly abated once we dig deeper into the mathematical structure underlying the process under study.[2] Unless it embraces the idea that humans are capable of eventually deciphering the main mathematical intricacies of the universe, this current implicitly assumes that the quest for larger and more descriptive models may not have an end.

If an anti-Platonic understanding of mathematics is equally defensible, however, we must concede that our mathematical representations of physical phenomena may simply be metaphors without any objective proof. The success of mathematics in describing these phenomena does not necessarily demonstrate its existence:[3] Field (2016) and Balaguer (1996) showed that Newtonian physics and quantum mechanics can be precisely nominalized and described without mathematics, pointing towards the scientific usefulness of mathematics as being mostly pragmatic, not idealistic—mathematics just makes calculations simpler. Whatever fit there is between a mathematical representation of a regularity and the regularity itself may hence occur in the mind of the observer and not in the real world. According to Lakoff and Núñez (2000: 81), this makes mathematics a human (and not divine) phenomenon:[4]

[1] This is especially apparent if one observes the evolution of climate, hydrological, and epidemiological models over the last 80 years. For instance, the simple general circulation models of the 1960s have become exhaustive atmosphere–ocean general circulation models (Hausfather et al. 2020, Sarofim et al. 2021). The classic bucket-type models of the 1970s have turned into global hydrological models that simulate the whole water cycle and the impact of humans in it (Manabe 1969, Bierkens 2015). In epidemiology, the first compartment models in the 1930s and 1940s had just a few parameters (Kermack and McKendrick 1933); the COVID-19 model of the Imperial College London had c.900 (Ferguson et al. 2020). See Puy et al. (2022) for further details.

[2] The design of finer-grained models is regarded as a critical step towards that aim. In climate prediction, more detailed models are assumed to solve convective cloud systems and their influence upon the reflection of incoming solar radiation (Palmer 2014, Schär et al. 2020, Fuhrer et al. 2018). In hydrology, finer-grained detail is thought to provide better representations of human water demands, land–atmospheric interactions, topography and vegetation and soil moisture, and evapotranspiration processes (Bierkens 2015, Wood et al. 2011). See Beven and Cloke (2012) and Beven et al. (2015) for a critique of the drive towards model hyper-resolution in hydrology.

[3] The assumption that there is a mathematical truth because mathematics is essential to explaining the success of science is known as the Quine–Putnam indispensability argument for mathematical realism (Putnam 1975), and has been undermined by, among others, Maddy (1992) and Sober (1993).

[4] The application of mathematics to the social sciences is also a subject of concern, both because it imposes possibly dangerous styles of thinking on the social issue under study and because as an element of education and regimentation, mathematics takes on a performative, non-neutral role (Ernest 2018). The mathematization of economics is lamented by Drechsler (2000), Mirowski (1989), Reinert (2019), and Romer (2015). See also Drechsler and Fuchs in this volume for a critique of a 'quantitative-mathematical social science'.

it follows from the empirical study of numbers as a product of mind that it is natural for people to believe that numbers are not a product of mind!

Mathematical models in this current would be meta-metaphors, given their higher-order abstraction of a set of already abstract concepts. By assuming the absence of transcendent mathematics, this view ultimately regards models as figurative representations of an allegory and thus as intrinsically uncertain constructs—their design, scope, and conceptualization is not defined by the underlying set of mechanisms modelled as much as by the subjective perspective of the modeller(s), which inevitably reflects their methodological and disciplinary configuration (Saltelli et al. 2020). The addition of model detail as a means to get closer to the 'truth' may lead a modeller astray in the pursuit of a 'truth' that only exists as such in the modeller's mind.

The predominance of the Platonic view in mathematical modelling is a distinctive tenet of modernity and its willingness to unfold universal, timeless physical truths over local, timely principles—what Toulmin (1990) refers to as the triumph of Descartes' certainty over Montaigne's doubt. The progressive formalization of reality into numbers and equations of increasing sophistication reflects the thirst for developing more objective, rational solutions to the social-environmental challenges that our society faces. A conspicuous example of this ethos is the Destination Earth Initiative launched in 2022 by the European Commission (2022), which aims to create a digital twin ('a highly accurate model') of the Earth to 'monitor, model and predict natural and human activity', ultimately helping to 'tackle climate change and protect nature'. The same can be said of the willingness of a section of the hydrological community to simulate global water flows at a 1 km resolution and every 1–3 h in order to 'adequately address critical water science questions' (Wood et al. 2011). This ambition resembles Carroll (1893)'s fictional tale of a nation that developed a map with a scale of a mile to the mile, or Borges' (1998) story of the map of the Empire whose 'size was that of the Empire, and which coincided point for point with it'. In both tales, the map ends up being discarded as useless.

Carrol's and Borges' tales do not differ much from what models are in the anti-Platonic view of mathematics, given their abstraction of ideas with a likeness to the empirical world. And yet their figurative meaning does not diminish their capacity to produce reflection and wisdom: we all get what the insight is and how the fictional stories connect with our experiences. The tales are not 'out there', but they convey an understanding about a complex, specific feature of human behaviour in a plain and simple way. The addition of narrative detail may or may not bring nuance to the story, and hence it is not an end in itself. If models can philosophically be thought of as meta-metaphors in a legitimate way, then their metaphysical status may be indistinguishable from that of any literary work—mathematical units such as π would be indistinguishable from fictional characters such as Sherlock Holmes (Bonevac 2009: 345).

The philosophical validity of the anti-Platonic view of mathematics opens the door to important questions with regard to our overreliance on the use of models as tools for prediction, management, and control. It also suggests caution in the quest for ever more complex models so as to achieve sharper insights. Our use of the Greek word 'hubris' in the title denotes the self-assured arrogance that comes with a lack of attention to these epistemological issues. For Jasanoff (2003: 238), mathematical models are 'technologies of hubris' when they aim to predict phenomena to facilitate management and control in areas of high uncertainty. The addition of conceptual depth to a model with unexplored ambiguities makes the hubris problem worse. In the next pages, we discuss four specific issues derived from this problem: explosion of the uncertainty space, black-boxing, computational exhaustion, and model attachment. We conclude by offering some reflections on why mathematical models may benefit from relaxing the pursuit of the Cartesian dream.

Explosion of the uncertainty space

In the environmental, climate, or epidemiological sciences, the addition of model detail often involves increasing the spatial/temporal resolution of the model, adding new parameters and feedbacks, or enlarging the model with new compartments linked in a causal chain: for instance, a model simulating how an aquifer gets polluted may be extended with compartments that model how this pollution reaches the surface, how it is metabolized by crops, and, finally, how it impacts the health of the affected population (Goodwin, Canada, and Saltelli 1987). During this model expansion process, new parameters need to be estimated with their error due to natural variation, measurement bias, or disagreement among experts about their 'true' value. There might also be ambiguity as to which is the best way to mathematically integrate the new parameters into the model. All this often makes the model output uncertainty increase, and not decrease, at each model upgrade stage. Scholars refer to this phenomenon as the 'uncertainty cascade' effect, where uncertainties add up and expand with model complexification (Hillerbrand 2014, Wilby and Dessai 2010).

Whydo more detailed mathematical models tend to produce more uncertain estimates? To understand this paradox, we draw attention to the concept of the model's *uncertainty space*, which is the space formed by the number of uncertain parameters k. When $k = 1$, the uncertainty space can be represented as a segment; when $k = 2$, as a plane; when $k = 3$, as a cube; and when $k > 3$, as a k-dimensional hypercube. Now let us assume that all parameters are distributed in the interval [0, 1]. Ten points suffice to sample the segment with a distance of 0.1 between points; if we want to keep the same sampling density to sample the plane, the cube, and the k-dimensional hypercube, we will need 100, 1000, and 10^k points

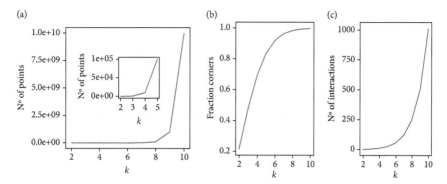

Fig. 4.1 The curse of dimensionality: (a) Increase in the number of points needed to evenly sample an uncertainty space with a distance of 0.1 between adjacent points as a function of the number of model parameters k; (b) Fraction of the space occupied by the corners of the hypercube along k dimensions; (c) Number of possible interactions between parameters as a function of k.

respectively. Note how the number of sampling points increases exponentially with every addition of a new parameter (Fig. 4.1a).

The more parameters we add in our model, the larger the proportion of the uncertainty space (and hence the number of sampling points) that will be located in the corners and edges of the hypercube. To illustrate this fact, let us think about the ratio of the volume of a hypersphere of radius ½ to the volume of a unit hypercube of dimension k (Saltelli and Annoni 2010). The centre of the uncertainty space would equal the hypersphere, whereas the corners would equal the space of the hypercube outside the hypersphere. For $k = 2$ and $k = 3$, these geometrical forms can be pictured as a circle inscribed in a plane and as a sphere within a cube. Consider how the volume of the space occupied by the corners increases exponentially with the addition of parameters: for $k = 2$ it amounts to 21% of the space (Fig. 4.2a); for $k = 3$ it corresponds to 47% of the space (Fig. 4.2b). By the time we reach $k = 10$, the corners already form 99.7% of the volume of the hypercube (Fig. 4.1b).

It is in the corners and edges of the hypercube where high-order effects, for example interactions between parameters whose effect on the model output uncertainty is larger than their individual effects (Saltelli et al. 2008), tend to occur. Non-additive operations such as multiplications, divisions, or exponentials are enough to promote interactions, whose number grows as $2^k - k - 1$. For a model with three parameters $f(\mathbf{x})$, $\mathbf{x} = (x_1, x_2, x_3)$, there might be four interactions up to the third order, for example three two-order interactions (x_1x_2, x_1x_3, x_2x_3) plus one three-order interaction $(x_1x_2x_3)$. The addition of one and two extra parameters raises the number of possible interactions to 11 and 26 and up to four- and five-order effects respectively. With 10 parameters, the number of possible interactions up to the 10th-order effect rises to 1024 (Fig. 4.1c).

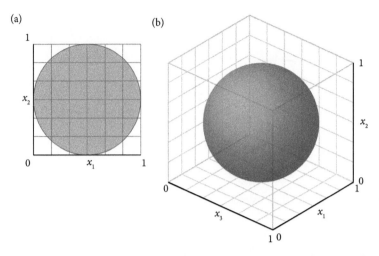

Fig. 4.2 Increase in the corners of the uncertainty space as a function of the number of parameters k: (a) $k = 2$, (b) $k = 3$.

This explosion of the uncertainty space with every addition of a parameter means that finer-grained models have uncertainty spaces with disproportionately larger corners and potential higher-order interactions, a consequence of the curse of dimensionality (Bellman 1957). If active, these high-order effects tend to boost the model output uncertainty. And the order of the highest-order effect active in the model tends to be higher in models with a larger number of uncertain parameters and structures (Puy et al. 2022). This explains why the addition of model detail may not necessarily increase accuracy, but rather swamp the model output in indeterminacy. This becomes apparent *if and only if* the model's uncertainty space is thoroughly explored. If it is not, its estimates will be spuriously accurate. And highly detailed models tend to be so computationally demanding to run that the exploration of their uncertainty space often becomes unaffordable.

Black-boxing

The accumulation of parameters, feedbacks, and interlinked compartments creates models whose behaviour may not be easy to grasp intuitively. It is customary for models in the climate, environmental, or epidemiological sciences to be formed by thousands of lines of code added up over decades and prone to include bugs, undesired behaviours, and deprecated or obsolete language features. Global climate models, for example, are on average composed of 500,000 lines of code in Fortran (with the Community Earth System Model (CESM) of the Department of Energy and the National Science Foundation peaking at *c*.1,300,000 lines of code)

(Fig. 4.3a), and include both obsolete statements and important cyclomatic complexities (Fig. 4.3b) (Méndez Tinetti, and Overbey 2014). The model of Imperial College London that underpinned the UK's response to COVID-19 had *c.*15,000 lines of code written over ~13 years, with several dangerous coding constructs, undefined behaviours, and violations of code conventions (Zapletal et al. 2021). If we consider that in the software development industry (whose quality control practices are more stringent than those of academia (Källén 2021)), there may be on average 1–25 errors per 1000 lines of code (McConnell 2004), we may get to appreciate the number of undetected errors potentially hiding in the code of big scientific models. Lubarsky's Law of Cybernetic Entomology ('There is always one more bug') looms larger in the quest for ever more exhaustive simulations.

Even under an optimistic scenario (for example, only 10% of the bugs are capable of meaningfully affecting the results without the analyst noticing; only 50% of the model is executed; there are 10 errors per 1000 lines of code), the chance of obtaining wrong results with large models, say with a model formed by 20,000 lines of code, may become a certainty (Soergel 2015).

To increase quality and open up the model's black box, several authors have requested modelling teams to give open source software licences to their codes and post them in repositories for inspection, error detection, and reuse (Morin et al. 2012, Barton et al. 2020). The addition of comments and a user's manual may also improve usability and help with pinpointing undesirable behaviours due to

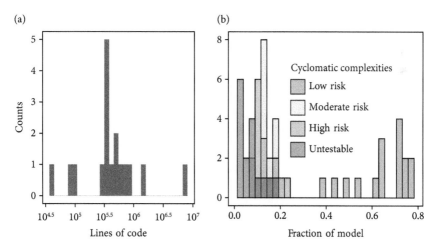

Fig. 4.3 Code of Global Climate Models (GCM): (a) Lines of source code; (b) Fraction of sub-programs in each GCM with cyclomatic complexities, defined as the number of independent paths through a program unit (McCabe 1976). Softwares with a higher cyclomatic complexity are harder to test and riskier to modify without introducing faults in the code, and are more prone to include bugs (Chowdhury and Zulkernine 2010). The plots are based on the data provided by Méndez, Tinetti, and Overbey (2014) for 16 GCM.

programming mistakes. These initiatives address the technical problems that arise when writing and upgrading code to improve the model's descriptive capacity, but fall short in handling another consequence of model hubris: the expansion in the number of value-laden assumptions embedded in the code (Saltelli et al. 2020a). Such features escape the formal checks used to locate and correct code defects, and yet their impact in the simulations can easily be substantial: a bug in the code may offset the output of an algorithm that informs hurricane forecasts by 30% (Perkel 2022); the assumption that farmers prioritize long- rather than short-maturing maize varieties can lead an algorithm to leave several million people without water insurance payouts (Johnson 2020).

A paradigmatic example of how the expansion of a model can be affected by the addition of potentially problematic assumptions can be found in the use of Integrated Assessment Models (IAMs) to guide policies against climate change. When the 1.5 °C goal became the target of climate action after the 2015 Paris Agreement, IAMs were upgraded with modules that allowed this target to be achieved by assuming a wide and intensive use of negative emission technologies (for example, bioenergy, carbon capture, and storage) (van Beek et al. 2022, Anderson and Peters 2016). Such a vision implied that a technology still in its infancy could be deployed to fight climate change on a global scale and deliver as expected 15 years later. Many other questionable assumptions are likely to hide behind IAMs—in fact, their number may be as large as to be unmanageable in the context of scientific publishing practices (Skea et al. 2021). This effectively places IAMs beyond the reach of peer reviews (Rosen 2015).

Computational exhaustion

Until 2004–6, the development of increasingly finer-grained models was fuelled by computational advances that allowed the number of transistors in a chip to be doubled while keeping the power requirements per unit of area constant. In other words: the speed of arithmetic operations could grow exponentially without significantly increasing power consumption. The doubling of the number of transistors and the scaling between chip area and power use are known as Moore's law and Dennard's scaling respectively (Moore 1965, Dennard et al. 1974), and are two key trends needed to understand the evolution of computing performance and mathematical modelling over the last 50 years. Among others, they facilitated the onset of our current numerical, physics-based approach to model planetary climate change by sustaining the development of a 'hierarchy of models of increasing complexity', as put by computer pioneer and meteorologist Jule G. Charney (Balaji 2021).

After 2006, it became apparent that this trend was over (Fig. 4.4). By this point, the doubling in the number of transistors in a chip could no longer be matched by a proportional boost in energy efficiency. This resulted in chips with increased power density and more heat being generated per unit of area, which had to be dissipated to prevent thermal runaway and breakdown. A solution was to keep a fraction of the chip power-gated or idled—the part known as 'dark silicon' (Taylor 2012). The higher the density in the number of transistors, the larger the fraction of the chip that has to be underused, a limitation that tapers off the improvement in clock frequencies and single-thread performances (Fig. 4.4a). Since higher chip densities are being required to match the demands of machine learning, big data, and mathematical models, the fraction occupied by 'dark silicon' may soon be larger than 90% (Taylor 2012, Kanduri et al. 2017) (Fig. 4.4b). Adding extra cores to improve performance is unlikely to sort this issue out, due to its meagre computational returns (Esmaeilzadeh et al. 2011).

Power consumption hence imposes a severe constraint on the addition of model resolution and suggests that, under our current computational paradigm, the quest towards ever more detailed models is computationally unsustainable. And the problem is one not only of arithmetic performance, but also of data storage. The 100 models participating in phase 6 of the Coupled Model Intercomparison Project currently produce c.80 petabytes of data;[5] if they increase resolution at ~1 km, the output may reach 45,000 PB (Schär et al. 2020), ~2,200 times more data than that stored in the Library of Congress in 2019 (~20 PB) (Spurlock 2019). To tackle the data avalanche derived from hyperresolution, some authors have

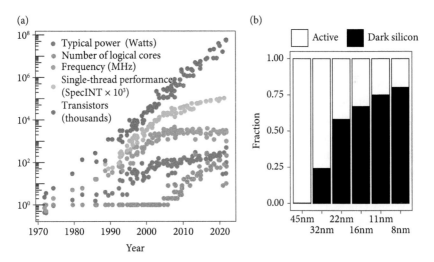

Fig. 4.4 Computational limits: (a) Microprocessor trend data up to the year 2020, retrieved from Karl Rupp in https://github.com/karlrupp/microprocessor-trend-data (last accessed 25 May 2022); (b) Fraction occupied by 'dark silicon' as a function of the technology (x-axis). Modified from Fig. 1.1 in Kanduri et al. (2017).

[5] 1 petabyte (PB) = 1000 terabytes (TB).

suggested to just save the simulation setup and re-run the simulation on demand (Schär et al. 2020), yet this strategy may breach the FAIR data principles whereby data should be findable, accessible, interoperable, and replicable (Wilkinson et al. 2016), and recomputing may be even more costly than archiving (Bauer et al. 2015).

Model attachment

The pursuit of finer-grained models requires the training of specialists able to set, calibrate, upgrade, and analyse these models and offer appropriate insights for scientific and policy purposes. The acquisition of this expertise is time-consuming and usually demands several years of education at a university or research institution. As with any other learning process, this situation sets the ground for the development of a domain-specific knowledge that endows its possessor with the capacity to efficiently put the learned skills to use in exchange for a higher risk of suffering cognitive biases such as Maslow (1966)'s hammer—'If the only tool you have is a hammer, it is tempting to treat everything as if it were a nail.' In mathematical modelling, this phenomenon takes place when a model is used repeatedly regardless of purpose or adequacy (a model of 'everything and everywhere'), a bias that has been attested for hydrological modelling: based on a sample size of c.1500 papers and seven global hydrological models (GHM), Addor and Melsen (2019) observed that in 74% of the studies, the model selected could be predicted based solely on the institutional affiliation of the first author.

Here we extend the work by Addor and Melsen (2019) to 13 new global (hydrological, vegetation, and land surface) models across c.1000 papers,[6] with the same results: the institutional affiliation of the first author anticipates the model used in ~75% of the publications (Fig. 4.5a). In fact, several institutions display a very high level of attachment. This is the case for the Institut national de recherche pour l'agriculture, l'alimentation et l'environnement (INRAE (National Research Institute for Agriculture, Food and the Environment)) (100%, GR4J, 15 papers), Osaka University (100%, H08, 11 papers), Utrecht University (91%, PCR-GLOBWB, 33 papers), the Potsdam Institut für Klimafolgenforschung (Potsdam Institute for Climate Impact Research) (93%, LPJmL, 54 papers), the University of Exeter (96%, JULES-W1, 23 papers), or Universités de Recherche Françaises (Udice) (96%, ORCHIDEE, 28 papers) (Fig. 4.6a). The attachment of each institution to its favourite model is also very consistent over time (Fig. 4.5b).

---

[6] CLM, CwatM, DBHM, GR4J, H08, JULES-W1, LPJmL, MATSIRO, MHM, MPI-HM, ORCHIDEE, PCR-GLOBWBW, and WaterGAP; see Puy (2022) for the methods and the references.

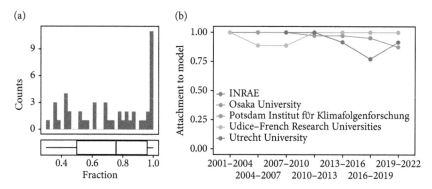

Fig. 4.5 Institutional attachment to global models: (a) Strength of the institutional attachment to a favourite model. A value of 1 means that 100 per cent of the studies published by a given institution rely on the same global model. Only institutions with more than five articles published are considered; (b) Strength of the attachment for some specific institutions through time.

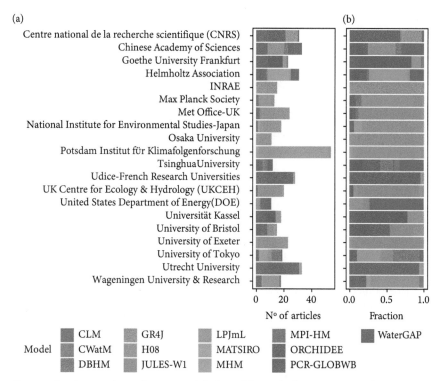

Fig. 4.6 Institutional attachment to global models. Only the 20 most productive institutions are displayed. (a) Stacked barplot with the number of publications per institution; (b) Fractioned stacked barplot.

The case of global models illustrates the extent to which model selection may be influenced by path dependencies that do not necessarily match criteria of adequacy, but rather convenience, experience, and habit (Addor and Melsen 2019). This inertia might be seen as an instance of the Einstellung effect, the insistence on using familiar tools and frames to tackle a problem even if they are suboptimal or inadequate (Luchins 1942). Very large models such as global models may become institutionalized to offset in the mid-term the costs derived from setting up a more efficient ecosystem in terms of writing code, training personnel, selecting modelling methods, calibrating algorithms, upgrading, and so on. Once a given modelling ecosystem is in place, the workflow is streamlined and the researcher has 'to do nothing but to press a few enters and then [the modelling pre-processing] is done' (Melsen 2022: 15). Switching to other models resets this process and moves the watch back for the institution and the researcher. Larger models may be more prone to promoting model attachment and become models of 'everything and everywhere', given the much higher costs involved in the first stages of their institutionalization (see also 'The model of everything: spoken and unspoken frames' in chapter 3).

Conclusion

In this chapter we have suggested moderating our Cartesian thirst to unfold the secrets of the natural world via increasingly detailed mathematical models. This pursuit may lead us to overlook important discussions on mathematical Platonism and the nature of mathematical knowledge. It may also promote black-boxing, computational exhaustion, and model attachment. There exists in mathematical modelling a 'political economy' whereby larger models command more epistemic authority and offer to their developer(s) better opportunities to defend them against criticisms. Thus the resistance of some modellers to coming to terms with the full uncertainty of their work may have motivations such as 'navigating the political' (van Beek et al. 2022)—that is, defending the role of modelling work in policy-relevant settings where epistemic authority needs to be preserved (Robertson 2021).

Modelling is but one among many instances of quantification where the issue of overambition, in the words of Quade (1980: 36), is a recurring theme. In that sense, sociology of quantification offers powerful instruments to dissect the normative implications of our ambitions to box reality behind numbers. There are instances of quantification where contestation becomes more natural, such as in the use of statistical indicators in socioeconomic domains, discounted by *statactivists* (Bruno, Didier, and Prévieux 2014, Mennicken and Salais 2022). Even when it comes to rating and ranking, there is a rich literature condemning the excesses of quantification (Muller 2018, O'Neil 2016), and a recent success of activism in this field is the fight against the Work Bank's Doing Business index (Cobham 2022) (see also chapter 9). The use of statistical rituals in sociology is observed by Gigerenzer and Marewski (2015), and the mathematization of the economy is discussed by Drechsler and Fuchs in chapter 6 of this book. As argued elsewhere (Saltelli 2019), mathematical modelling tends to be shielded from more stringent forms of critical activism by the barrier that the complexity of the mathematical constructs and coding pose to experts and non-experts alike.

Data availability

The code to reproduce the results and the figures of this study can be found in Puy (2022) and in GitHub (https://github.com/arnaldpuy/model_hubris).

References

Addor, N. and L.A. Melsen. 2019. 'Legacy, Rather than Adequacy, Drives the Selection of Hydrological Models'. *Water Resources Research*, 55(1), 378–90, doi: 10.1029/2018WR022958.

Anderson, K. and G. Peters. 2016. 'The Trouble with Negative Emissions'. *Science*, 354(6309), 182–3, doi: 10.1126/science.aah4567.

Balaguer, M. 1996. 'Towards a Nominalization of Quantum Mechanics'. *Mind*, 105(418), 209–26.

Balaguer, M. 1998. *Platonism and Anti-Platonism in Mathematics*, Oxford University Press, Oxford.

Balaji, V. 2021. 'Climbing Down Charney's Ladder: Machine Learning and the Post-Dennard Era of Computational Climate Science'. *Philosophical Transactions of the Royal Society A: Mathematical, Physical and Engineering Sciences*, 379, 20200085, doi: 10.1098/rsta.2020.0085.

Barton, C.M. et al. 2020. 'Call for Transparency of COVID-19 Models'. *Science*, 368(6490), edited by J. Sills, 482–3, doi: 10.1126/science.abb8637.

Bauer, P., A. Thorpe, and G. Brunet. 2015. 'The Quiet Revolution of Numerical Weather Prediction'. *Nature*, 525(7567), 47–55, doi: 10.1038/nature14956.

van Beek, L., J. Oomen, M. Hajer, P. Pelzer, and D. van Vuuren. 2022. 'Navigating the Political: An Analysis of Political Calibration of Integrated Assessment Modelling in Light of the 1.5 °C Goal'. *Environmental Science & Policy*, 133, 193–202, doi: 10.1016/j.envsci.2022.03.024.

Bellman, R.E. 1957. *Dynamic Programming*, Princeton University Press, Princeton, NJ.

Benacerraf, P. 1973. 'Mathematical Truth'. *The Journal of Philosophy*, 70(19), 661–79.

Beven, K. and H. Cloke. 2012. 'Comment on "Hyperresolution Global Land Surface Modeling: Meeting a Grand Challenge for Monitoring Earth's Terrestrial Water" by Eric F. Wood et al.' *Water Resources Research*, 48(1), 2–4, doi: 10.1029/2011WR010982.

Beven, K., H. Cloke, F. Pappenberger, R. Lamb, and N. Hunter. 2015. 'Hyperresolution Information and Hyperresolution Ignorance in Modelling the Hydrology of the Land Surface'. *Science China Earth Sciences*, 58(1), 25–35, doi: 10.1007/s11430-014-5003-4.

Bierkens, M.F.P. 2015. 'Global Hydrology 2015: State, Trends, and Directions'. *Water Resources Research*, 51(7), 4923–47, doi: 10.1002/2015WR017173.

Bonevac, D. 2009. 'Fictionalism'. In *Philosophy of Mathematics*, edited by A. Irvine, Elsevier, North Holland, pp. 345–93.

Borges, J.L. 1998. 'On Exactitude in Science'. In *Jorge Luis Borges, Collected, Fictions*, translated by H. Hurley, Penguin Books, London, p. 325.

Bruno, I., E. Didier, and J. Prévieux, eds. 2014. *Statactivisme. Comment Lutter Avec Des Nombres*, La Découverte, Paris.

Carroll, L. 1893. *Sylvie and Bruno Concluded*. Macmillan and Co, London.

Chowdhury, I. and M. Zulkernine. 2010. 'Can Complexity, Coupling, and Cohesion Metrics Be Used as Early Indicators of Vulnerabilities?' *SAC '10: Proceedings of the 2010 ACM Symposium on Applied Computing*, 1963–9, doi: 10.1145/1774088.1774504.

Cobham, A. 2022. 'What the BEEP? The World Bank Is Doing Business Again'. *Tax Justice Network*, 17 March 2022. https://taxjustice.net/2022/03/17/what-the-beep-the-world-bank-is-doing-business-again/.

Davis, P.J. and R. Hersh. 1986. *Descartes' Dream: The World According to Mathematics*, Dover Publications, Mineola, NY.

Dennard, R., F. Gaensslen, H.-N. Yu, V. Rideout, E. Bassous, and A. LeBlanc. 1974. 'Design of Ion-Implanted MOSFET's with Very Small Physical Dimensions'. *IEEE Journal of Solid-State Circuits*, 9(5), 256–68, doi: 10.1109/JSSC.1974.1050511.

Descartes, R. 2006. *A Discourse on the Method*, edited by I. Maclean, Oxford University Press, Oxford.

Drechsler, W. 2000. 'On the Possibility of Quantitative-Mathematical Social Science, Chiefly Economics'. *Journal of Economic Studies*, 27(4/5), 246–59, doi: 10.1108/01443580010341727.

Ernest, P. 2018. 'The Ethics of Mathematics: Is Mathematics Harmful?' In *The Philosophy of Mathematics Education Today*, edited by P. Ernest, Springer International Publishing, Cham, pp. 187–218.

Esmaeilzadeh, H., E. Blem, R.S. Amant, K. Sankaralingam, and D. Burger. 2011. 'Dark Silicon and the End of Multicore Scaling', *2011 38th Annual International Symposium on Computer Architecture (ISCA)*, IEEE, 365–76.

European Commission. 2022. *Destination Earth – New Digital Twin of the Earth Will Help Tackle Climate Change and Protect Nature*. https://ec.europa.eu/commission/presscorner/detail/en/ip_22_1977.

Ferguson, N.M. et al. 2020. *Report 9: Impact of Non-Pharmaceutical Interventions (NPIs) to Reduce COVID-19 Mortality and Healthcare Demand*, Tech. rep. March, 1–20.

Field, H. 1989. *Realism, Mathematics and Modality*, Blackwell, Hoboken, NJ.

Field, H. 2016. *Science Without Numbers: A Defence of Nominalism*, 2nd edition, Oxford University Press, Oxford.

Frege, G. 2007. *Gottlob Frege: Foundations of Arithmetic (Longman Library of Primary Sources in Philosophy)*, edited by D. Kolak, 1st edition, Routledge, Abingdon.

Fuhrer, O., T. Chadha, T. Hoefler, G. Kwasniewski, X. Lapillonne, D. Leutwyler, D. Lüthi, C. Osuna, C. Schär, T.C. Schulthess, and H. Vogt. 2018. 'Near-Global Climate Simulation at 1 km Resolution: Establishing a Performance Baseline on 4888 GPUs with COSMO 5.0'. *Geoscientific Model Development*, 11(4), 1665–81, doi: 10.5194/gmd-11-1665-2018.

Funtowicz, S. and Â.G. Pereira. 2015. 'Cartesian Dreams'. In *Science, Philosophy and Sustainability. The End of the Cartesian Dream*, edited by Â.G. Pereira and S. Funtowicz, Routledge, Abingdon, pp. 1–9.

Gigerenzer, G. and J.N. Marewski. 2015. 'Surrogate Science: The Idol of a Universal Method for Scientific Inference'. *Journal of Management*, 41(2), 421–40, doi: 10. 1177/0149206314547522.

Gödel, K. 1995. 'Some Basic Theorems on the Foundations of Mathematics and Their Implications'. *Collected Works. Volume III. Unpublished Essays and Lectures*, edited by S. Feferman, J. W. D. Jr, W. Goldfarb, C. Parsons, and R. N. Solovay, Oxford University Press, Oxford, pp. 304–24.

Goodwin, B.W., A.E.C.L. Canada, and A. Saltelli. 1987. *PSACOIN Level 0 Intercomparison*. Tech. Rep. Nuclear Energy Agency, Organisation for Economic Co-operation and Development, Paris.

Hausfather, Z., H.F. Drake, T. Abbott, and G.A. Schmidt. 2020. 'Evaluating the Performance of Past Climate Model Projections'. *Geophysical Research Letters*, 47(1), 1–10, doi: 10.1029/2019GL085378.

Hillerbrand, R. 2014. 'Climate Simulations: Uncertain Projections for an Uncertain World'. *Journal for General Philosophy of Science*, 45(1), 17–32, doi: 10.1007/s10838-014- 9266-4.

Jasanoff, S. 2003. 'Technologies of Humility: Citizen Participation in Governing Science'. *Minerva*, 41, 223–44, doi: 10.1023/A:1025557512320.

Johnson, L. 2020. 'Sharing Risks or Proliferating Uncertainties? Insurance, Disaster and Development'. In *The Politics of Uncertainty. Challenges of Transformation*, edited by I. Scoones and A. Stirling, Routledge, Abingdon, pp. 45–57.

Källén, M. 2021. 'Towards Higher Code Quality in Scientific Computing', PhD thesis, University of Uppsala.

Kanduri, A., A.M. Rahmani, P. Liljeberg, A. Hemani, A. Jantsch, and H. Tenhunen. 2017. 'A Perspective on Dark Silicon', Springer International Publishing, Cham, pp. 3–20, doi: 10.1007/978-3-319-31596-6_1.

Kermack, W.O. and A.G. McKendrick. 1933. 'Contributions to the Mathematical Theory of Epidemics. III.—Further Studies of the Problem of Endemicity'. *Proceedings of the Royal Society of London. Series A, Containing Papers of a Mathematical and Physical Character*, 141(843), 94–122, doi: 10.1098/rspa.1933.0106.

Lakoff, G. and R.E. Núñez. 2000. *Where Mathematics Comes From: How the Embodied Mind Brings Mathematics into Being*, Basic Books, New York.

Luchins, A.S. 1942. 'Mechanization in Problem Solving: The Effect of Einstellung'. *Psychological Monographs*, 54(6), i–95, doi: 10.1037/h0093502.

Maddy, P. 1992. 'Indispensability and Practice'. *The Journal of Philosophy*, 89(6), 275–89.

Manabe, S. 1969. 'Climate and the Ocean Circulation 1: I. The Atmospheric Circulation and the Hydrology of the Earth's Surface'. *Monthly Weather Review*, 97(11), 739–74, doi: 10.1175/1520-0493(1969)097<0739:catoc>2.3.CO;2.

Maslow, A.H. 1966. *The Psychology of Science: A Reconnaissance*, Gateway Editions, South Bend, IN.

McCabe, T.J. 1976. 'A Complexity Measure'. *IEEE Transactions on Software Engineering SE-2(4)*, 308–20, doi: 10.1109/TSE.1976.233837.

McConnell, S. 2004. *Code Complete, Second Edition*, Microsoft Press, Redmond, WA.

Melsen, L.A. 2022. 'It Takes a Village to Run a Model—The Social Practices of Hydrological Modeling'. *Water Resources Research, 58(2)*, e2021WR030600, doi: 10.1029/2021WR030600.

Méndez, M., F.G. Tinetti, and J.L. Overbey. 2014. 'Climate Models: Challenges for Fortran Development Tools'. *IEEE*, 6–12, doi: 10.1109/SE-HPCCSE.2014.7.

Mennicken, A. and R. Salais, eds. 2022. *The New Politics of Numbers. Utopia, Evidence and Democracy*, Palgrave Macmillan, London.

Mirowski, P. 1989. *More Heat than Light: Economic as Social Physics, Physics as Nature's Economics*, Cambridge University Press, Cambridge, doi: 10.1017/CBO9780511559990.

Moore, G.E. 1965. 'Cramming More Components onto Integrated Circuits'. *Electronics Magazine, 38*, 114–17.

Morin, A., J. Urban, P.D. Adams, I. Foster, A. Sali, D. Baker, and P. Sliz. 2012. 'Shining Light into Black Boxes'. *Science, 336(6078)*, 159–60, doi: 10.1126/science.1218263.

Muller, J. Z. 2018. *The Tyranny of Metrics*, Princeton University Press, Princeton, NJ.

O'Neil, C. 2016. *Weapons of Math Destruction: How Big Data Increases Inequality and Threatens Democracy*, Penguin Random House, London.

Palmer, T. 2014. 'Climate Forecasting: Build High-Resolution Global Climate Models'. *Nature, 515(7527)*, 338–9, doi: 10.1038/515338a.

Perkel, J.M. 2022. 'How to Fix Your Scientific Coding Errors'. *Nature, 602(7895)*, 172–3, doi: 10.1038/d41586-022-00217-0.

Putnam, H. 1975. 'What Is Mathematical Truth?' In *Mathematics Matter and Method: Philosophical Papers, Volume 1*, 1st edition, Cambridge University Press, Cambridge, pp. 60–78.

Puy, A. 2022. 'R Code of Mind the Hubris in Mathematical Modeling'. *Zenodo*, doi: 10.5281/zenodo.6598848.

Puy, A., P. Beneventano, S.A. Levin, S.L. Piano, T. Portaluri, and A. Saltelli. 2022. 'Models with Higher Effective Dimensions Tend to Produce More Uncertain Estimates'. *Science Advances, 8(42)*, eabn9450, doi: 10.1126/sciadv.abn9450.

Quade, E.S. 1980. 'Pitfalls in Formulation and Modeling'. In *Pitfalls of Analysis*, edited by G. Majone and E.S. Quade, John Wiley & Sons, Chichester, UK, pp. 23–43.

Reinert, E.S. 2019. 'Full Circle: Economics from Scholasticism through Innovation and Back into Mathematical Scholasticism: Reflections on a 1769 Price Essay: "Why Is It That Economics So Far Has Gained So Few Advantages from Physics and Mathemat-ics?"' In *The Visionary Realism of German Economics: From the Thirty Years' War to the Cold War*, edited by R. Kattel, Anthem Press, London, 555–68.

Robertson, S. 2021. 'Transparency, Trust, and Integrated Assessment Models: An Ethical Consideration for the Intergovernmental Panel on Climate Change'. *WIREs Climate Change, 12(1)*, doi: 10.1002/wcc.679.

Romer, P. 2015. 'Feynman Integrity'. *Paul Romer*, 1 August 2015. https://paulromer.net/feynman-integrity/ (last accessed 23 March 2023).

Rosen, R.A. 2015. 'IAMs and Peer Review'. *Nature Climate Change, 5(5)*, 390, doi: 10.1038/nclimate2582.

Saltelli, A. 2019. 'A Short Comment on Statistical versus Mathematical Modelling'. *Nature Communications, 10(1)*, 8–10, doi: 10.1038/s41467-019-11865-8.

Saltelli, A. and P. Annoni. 2010. 'How to Avoid a Perfunctory Sensitivity Analysis'. *Environmental Modelling and Software, 25(12)*, 1508–17, doi: 10.1016/j.envsoft.2010. 04.012.

Saltelli, A., G. Bammer, I. Bruno, E. Charters, M. D. Fiore, E. Didier, W.N. Espeland, J. Kay, S.L. Piano, D. Mayo, R.P. Jr, T. Portaluri, T.M. Porter, A. Puy, I. Rafols, J.R. Ravetz, E. Reinert, D. Sarewitz, P.B. Stark, A. Stirling, J. van der Sluijs, and P. Vineis. 2020a. 'Five Ways to Ensure That Models Serve Society: A Manifesto'. *Nature, 582(7813)*, 482–4, doi: 10.1038/d41586-020-01812-9.

Saltelli, A., L. Benini, S. Funtowicz, M. Giampietro, M. Kaiser, E. Reinert, and J.P. van der Sluijs. 2020b. 'The Technique Is Never Neutral: How Methodological Choices Condition the Generation of Narratives for Sustainability'. *Environmental Science & Policy, 106*, 87–98, doi: 10.1016/j.envsci.2020.01.008.

Saltelli, A., M. Ratto, T. Andres, F. Campolongo, J. Cariboni, D. Gatelli, M. Saisana, and S. Tarantola. 2008. *Global Sensitivity Analysis. The Primer*, John Wiley & Sons, Chichester, UK, doi: 10.1002/9780470725184.

Sarofim, M.C., J.B. Smith, A. St Juliana, and C. Hartin. 2021. 'Improving Reduced Complexity Model Assessment and Usability'. *Nature Climate Change, 11*, 9–11, doi: 10.1038/s41558-020-00973-9.

Schär, C., O. Fuhrer, A. Arteaga, N. Ban, C. Charpilloz, S.D. Girolamo, L. Hentgen, T. Hoefler, X. Lapillonne, D. Leutwyler, K. Osterried, D. Panosetti, S. Rüdisühli, L. Schlemmer, T.C. Schulthess, M. Sprenger, S. Ubbiali, and H. Wernli. 2020. 'Kilometer-Scale Climate Models: Prospects and Challenges'. *Bulletin of the American Meteorological Society, 101(5)*, E567–E587, doi: 10.1175/BAMS-D-18-0167.1.

Skea, J., P. Shukla, A.A. Khourdajie, and D. McCollum. 2021. 'Intergovernmental Panel on Climate Change: Transparency and Integrated Assessment Modeling'. *Wiley Interdisciplinary Reviews: Climate Change, 12(5)*, doi: 10.1002/wcc.727.

Sober, E. 1993. 'Mathematics and Indispensability'. *The Philosophical Review, 102(1)*, 35–57.

Soergel, D. 2015. 'Rampant Software Errors May Undermine Scientific Results [Version 2; Peer Review: 2 Approved]'. *F1000Research* 3.303, doi: 10.12688/f1000research. 5930.2.

Spurlock, R. 2019. ' Petabyte—How Much Information Could It Actually Hold?' *Cobalt Iron*, 31 October 2019. https://info.cobaltiron.com/blog/petabyte-how-much-information-could-it-actually-hold (last accessed 23 March 2023).

Taylor, M.B. 2012. 'Is Dark Silicon Useful?' *DAC '12: Proceedings of the 49th Annual Design Automation Conference*, ACM Press, 1131–6, doi: 10.1145/2228360.2228567.

Toulmin, S. 1990. *Cosmopolis: The Hidden Agenda of Modernity*, The University of Chicago Press, Chicago, IL.

Wilby, R.L. and S. Dessai. 2010. 'Robust Adaptation to Climate Change'. *Weather, 65(7)*, 180–5, doi: 10.1002/wea.543.

Wilkinson, M.D. et al. 2016. 'The FAIR Guiding Principles for Scientific Data Management and Stewardship'. *Scientific Data, 3*, 160018, doi: 10.1038/sdata.2016.18.

Wood, E.F., J.K. Roundy, T.J. Troy, L.P.H. van Beek, M.F.P. Bierkens, E. Blyth, A. de Roo, P. Döll, M. Ek, J. Famiglietti, D. Gochis, N. van de Giesen, P. Houser, P.R. Jaffé, S. Kollet, B. Lehner, D.P. Lettenmaier, C. Peters-Lidard, M. Sivapalan,

J. Sheffield, A. Wade, and P. Whitehead. 2011. 'Hyperresolution Global Land Surface Modeling: Meeting a Grand Challenge for Monitoring Earth's Terrestrial Water'. *Water Resources Research*, *47*(5), 1–10, doi: 10.1029/2010WR010090.

Zapletal, A., D. Höhler, C. Sinz, and A. Stamatakis. 2021. 'The SoftWipe Tool and Benchmark for Assessing Coding Standards Adherence of Scientific Software'. *Scientific Reports*, *11*(1), doi: 10.1038/s41598-021-89495-8.

5

Mind the assumptions

Quantify uncertainty and assess sensitivity

Emanuele Borgonovo

Assumptions are a fundamental pillar of any scientific investigation. There is no theory or modelling exercise that can live without assumptions. A search of the term 'assumption' in the *Stanford Encyclopedia of Philosophy* does not find a dedicated entry but instead finds the presence of this term in 839 entries that include 'scientific method', 'scientific discovery', and 'scientific logic'. Specifically, assumptions play a fundamental role in scientific modelling and scientific simulations. What, then, are assumptions? Is there a way in which we can converge on a definition of the term? While this is certainly out of the reach of this chapter, we will try to provide some initial reflections and insights.

According to the *Cambridge Dictionary*, an assumption is 'something that you accept as true without question or proof'. In a scientific modelling exercise, an assumption is a statement that establishes the value of a parameter or takes away elements that may not be of concern. Let us refer for a second to the Archimedean model of the lever (see Fig. 5.1).

Its typical graphical representation is composed of two lines, possibly one triangle and two arrows, representing, respectively, the effort, the load arm, the pivotal point, and the forces. This model was first conceived in the opus *On the Equilibrium of Planes* by Archimedes and has, since then, been fundamental in statics. The model is, for instance, at the basis of the trolleys we use in airports. However, Archimedes does not consider the type of material, the colour, or the brand of the trolley. These aspects have a relevance in marketing the trolley, but not in its physics. The discussion of Frigg and Hartmann (2020) is illuminating in this respect: assumptions help to strip away inessential elements from scientific models.

On the other hand, we can regard assumptions as the set of hypotheses (or statements) that need to hold before we can substantiate the use of a given scientific model. In other words; suppose that in the problem at hand, a set of conditions has to be realized simultaneously in order for a certain equation to become applicable; then an equation with a known form would be adequate to simulate the system or phenomenon at hand. Here, the definition of adequacy would need a

Emanuele Borgonovo, *Mind the assumptions*. In: *The politics of modelling.* Edited by: Andrea Saltelli and Monica Di Fiore, Oxford University Press. © Oxford University Press (2023). DOI: 10.1093/oso/9780198872412.003.0005

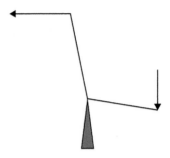

Fig. 5.1 A possible representation of the Archimedean model of the lever. We draw the effort arm, the force arm, the pivot, and the two forces at their extremes. The model does not have any reference to the colour or the material of the lever, but assumptions strip away irrelevant elements.

separate chapter, and we refer to Frigg and Hartmann (2020) and Winsberg (2019) for a thorough treatment. We content ourselves with an illustration: Einstein's prediction of the deviation of light is based on the theoretical principles of general relativity. These principles yield the mathematical equation from which to calculate the deviation. The light indeed undergoes the deflection forecast by the model, as physical experiments indicate. We have all three corners of the scientific triangle: a theory (first vertex) that produces a mathematical model (second vertex) that actually predicts a real-world phenomenon.

However, in several scientific investigations, we do not have all three vertices: scientists build mathematical models that try and reproduce a given phenomenon, often before a scientific theory is developed. In some cases, it will be impossible to achieve a complete scientific theory with axioms from which propositions are derived and that, in turn, yield the equations of the phenomenon at hand. This is likely the case in artificial intelligence investigations where a complete theoretical background about the phenomenon of interest (think of the case in which data come from social networks) is likely to remain out of reach. Under these conditions, we then encounter all those assumptions that a modeller makes in creating a simulation: these range from the selection of the mathematical form of the equations to the value of inputs and parameters.

We are then dealing with the vast class of models represented by computer simulators. Simulators help the scientist to obtain a numerical portrait of a system or phenomenon of interest.

Equations in computer codes aid engineers, physicists, and medical doctors with a variety of tasks. As defined in Winsberg (2019), we usually divide simulators into two main categories: equation-based simulators and agent-based simulators. An outstanding example of the former type of model is the simulator of the human heart developed by Alfio Quarteroni and his collaborators in a series of mathematical works (Quarteroni, Manzoni, and Vergara 2017). An example of the second model is the agent-based model of the financial market developed by Le Baron (2006).

Assumptions impact all aspects of scientific modelling. For instance, assumptions are at the basis of the theoretical principles that govern the model development itself. At the same time, they concern the values we attribute to parameters entering the final equations of the model. In a sense, it may well be useful to make a classification of assumptions, depending on the model element to which they refer. Assumptions can refer to the theoretical foundations of a model. We can call these assumptions *principles*, to give them a stronger connotation (Fig. 5.2). Principles are the conceptual basis of a scientific model or investigation: they comprise axioms and hypotheses of an underlying theory that grounds the model. As noted in Borgonovo et al. (2020), 'principles do not include specific algorithmic implementations; rather, they are conceptual guidelines that influence the modeler in formulating specific procedures or in choosing certain parameters'.

Regarding principles, the distinction between models of phenomena and models of data becomes relevant. In models of data, analysts 'assume' a form for the input–output mapping, typically selecting a parametric family and then calibrating the model using available data. This assumption is not necessarily dictated by a theory supporting the model. This is referred to as an 'absence of theory' by Begoli, Bhattacharya, and Kusnezov (2019), and is one of the main reasons for concerns about the reliability of results in machine learning models. Instead, researchers develop models of phenomena by following theoretical principles, and if the assumptions concerning the equations being formulated are verified, then these equations will produce an adequate description of the system behaviour.

To give tangible examples, let us consider two representative works, one in the realm of risk assessment (Apostolakis 1990) and the other in the business realm. These works deal with models developed to forecast the reliability of complex technological systems. Apostolakis introduces the notion of 'Model of the World' (MOW) to denote the system of equations that model the phenomenon

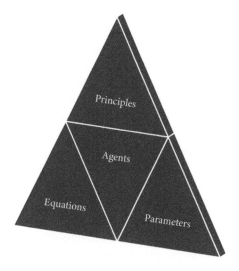

Fig. 5.2 A non-exhaustive set of elements of a scientific modelling problem.

of interest—of course, we should read this as the *model of the world of interest* in the specific investigation. To illustrate the notion, Apostolakis uses the Darcy equation for groundwater flow in saturated media:

$$q = -K\frac{\partial h}{\partial x},$$

where q is the specific discharge in the *x* direction, *h* is the hydraulic head, and K is the hydraulic conductivity. Ford W. Harris introduced the economic order quantity (EOQ) model (Harris 1913), which is a cornerstone of operations research. Harris's problem is that of determining the optimal lot size in a production system. While his work is written for a business magazine, his approach is strikingly scientific. Harris first states a sharp list of assumptions that clearly draw the reader to the heart of his problem. For instance, he assumes the *movement is regular* (Harris 1913: 948) (that is, stationarity). He then formulates an objective function (a total cost) based on the stated elements of the problem, and derives the now famous EOQ formula from a minimization:

$$Q = \sqrt{\frac{240MS}{C}}, \tag{1}$$

where Q is the optimal order quantity, S is the setup cost of an order, M is the monthly demand ('the movement of the stock' (Harris 1913: 947)), and C is the unit price of items in the stock. Under Harris's setting, this is the quantity that a rational manager should order. Nonetheless, Harris himself recognized the limitations of the model and warns us to use the numbers carefully (Cárdenas-Barrón, Chung, and Treviño-Garza 2014): 'But in deciding on the best size of order, the man responsible should consider all the factors that are mentioned [...] Hence, using the formula as a check, is at least warranted' (Harris 1913: 947).

Such a rule calls for a careful interpretation of any numerical indication obtained by models. This warning matches recent fears uttered in the scientific community. Examples are regarding the ethical guidelines in scientific modelling of the *Manifesto* of Saltelli et al. (2020), or the concerns related to the use (or abuse) of machine learning methods in Rudin (2019) and Begoli, Bhattacharya, and Kusnezov (2019).

These concerns lead us to the application of methods that can perform a proper uncertainty quantification of the results of the model. Let us take a step back and regard equation 5.1 as a black-box model in which three factors, namely, C, M, and S are fed into in a simulator that produces Q. Clearly, if we fix C, M, and S at nominal values C^0, M^0, and S^0, we have an optimal order quantity, Q^0.

At a level closer to the practice of modelling, besides assumptions about the characteristics of the system—that is, about stationarity or the fact that agents behave rationally—we have assumptions that regard parameters. As discussed in Borgonovo et al. (2020), parameters 'are cardinal quantities that influence the evolution of the model but are determined outside of the simulation run'.

In general, one regards the model as a black box, in which a set of inputs X is fed into a simulator $g(X)$ to produce an output Y of interest (in our case it would be $X = [C, M, S]$) (Fig. 5.3).

Consider now an analyst needing (or willing) to determine the EOQ, and whose available information allows her to set the following values for the inputs: $M^0 = 4500$, $S^0 = 3.5$, $C^0 = 75$. The corresponding EOQ calculated by the model is $Q = 224.5$ units. Note that if we were to commute this quantity to management, we would be exposed to the risk of acritically accepting the results of the model. Reporting a point value in a context full of uncertainties is evidently poor scientific practice (see the recent *Manifesto* of Saltelli et al. (2020)). In his work that established the foundations of sensitivity analysis in Economics, Paul Samuelson, a winner of the Nobel Prize in this field, writes: 'If no more than this could be said, the economist would be truly vulnerable to the gibe that he is only a parrot taught to say "supply and demand"' (Samuelson 1947: 97). The risk is one of undermining the entire effort of the modelling exercise.

In this context, we are interested in the variation of the model output that follows the variations in the inputs. This task comprises two important sub-tasks: sensitivity analysis and uncertainty quantification.

Regarding uncertainty quantification, we refer to the monographs of Saltelli et al. (2008), Sullivan (2015), and Borgonovo (2017), as a complete treatment is out of our reach in this chapter. However, let us briefly review the notion and procedure. The purpose of an uncertainty quantification is to assess the variability of the output of a model. If the model output is numerical, this variability can be expressed in the form of a variation range and, if possible, of a probability distribution of the model output over this range. These quantities are obtained by propagating the uncertainty in the model inputs all the way to the output. To illustrate, the analyst assigns plausible ranges to the inputs and, if possible, a corresponding probability distribution. Uncertainty is then propagated through the

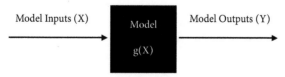

Fig. 5.3 The black-box view of the input–output mapping of a computer model.

model via a Monte Carlo simulation.[1] Suppose the scientist dealing with the Harris EOQ model sets the following ranges: $M \in [2000, 7000]$, $S \in [2, 5]$, $C \in [50, 100]$ and assigns uniform distributions to the inputs over these ranges. Monte Carlo propagation would lead to the distribution of Q, whose density is reported in Graph a) of Fig. 5.4.

Graph a) in Fig. 5.4 shows that the EOQ varies between 100 and 400 units. It is then up to the analyst or the manager to understand whether such variation is reasonable enough to proceed with the order, or if new analysis needs to be carried out to reduce the variability.

In general, the derivation of insights from sensitivity analysis can be made formal by the notion of sensitivity analysis setting. A setting is 'a way of framing the sensitivity quest in such a way that the answer can be confidently entrusted to a well-identified measure' (Saltelli et al. 2008: 24). In the case of Fig. 5.4, we call the setting a *stability setting*. This setting indeed comprises all those situations in which analysts wish to determine the robustness of the indications they are providing based on the numerical output of the model. For instance, consider an analyst dealing with a linear programming problem. Here, the range of variation in the model inputs over which the optimal plan remains the same is important to the modeller. A specific method that helps the analyst answer this question,

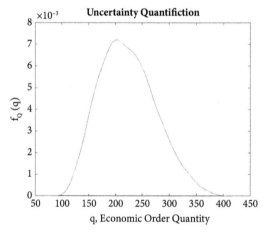

Fig. 5.4 Model output density, as a possible visualization of the outcome of the uncertainty quantification for our economic order quantity (EOQ) model in equation 5.1.

[1] For a characterization of Monte Carlo uncertainty propagation, please refer to Metropolis (1987) for a historical account, Glasserman (2003) for a technical treatment, and Winsberg (2019) for the epistemological viewpoint.

besides the uncertainty propagation we have seen, is Wendell's tolerance sensitivity (Wendell 1985).[2] Similarly, in a decision analysis problem expressed in the form of a decision tree, an influence diagram, or even a generic optimization problem, we find several techniques for assessing stability. One in particular is the value of information (Oakley 2009). If the value of information of a certain input is null, then the analyst can conclude that the optimal plan is insensitive to variations in that input.

A second task with which analysts are (or should be) frequently concerned is the answer to the question: how does the response of the model depend on each input marginally? Formally, we could call this setting *marginal behaviour determination*. The machine learning and the simulation literature has made available several tools to answer this question, especially in association with the urge to increase interpretability of machine findings. Much attention has been given to, and a number of advancements have been made in, partial dependence indicators, and we can point the reader towards the works of Hastie, Tibshirani, and Friedman (2001), Goldstein et al. (2015), and Apley and Zhu (2020) for further details.

In this chapter, we content ourselves with a visual interpretation: consider Fig. 5.5.

. The graphs in Fig. 5.5 immediately suggest to us that the EOQ is increasing in M and S and decreasing in C. The analyst can then infer answers to questions such as: is this behaviour consistent with intuition? What should we suggest to a decision maker based on this behaviour? Answering the first question also provides a way to enter into a debugging mode. If an underlying theory suggests that a model response should be decreasing in a given input and the sensitivity analysis suggests the opposite, it needs further investigation. We might either be on the verge of a scientific discovery in which a common (mis)conception is falsified or, more simply, there might be a bug, with a minus instead of a plus placed at some point of the (usually complex) code.

The tool displayed in Fig. 5.5 is one of the possible choices analysts have to determine marginal behaviour. Analysts can count on local approaches based on either differentiation (Griewank 2000), the sign of Newton ratios (Rakovec et al. 2014), or one-way sensitivity functions (van der Gaag, Renooij, and Coupe 2007). Global approaches comprise methods such as partial dependence functions (Hastie, Tibshirani, and Friedman 2001) and individual conditional expectation plots (Goldstein et al. 2015), as well as the recently introduced accumulated local effect plots (Apley and Zhu 2020). A review and critical analysis of these methods from a decision-making viewpoint was given recently in Borgonovo et al. (2021).

[2] Tolerance sensitivity is an approach tailored to linear programs. By exploiting the geometric properties of linear optimization, tolerance sensitivity determines the range of simultaneous variation in the inputs such that the optimal solution of the linear program remains stable. The inputs are the objective function coefficients, the right-hand sides, or the elements of the coefficient matrix.

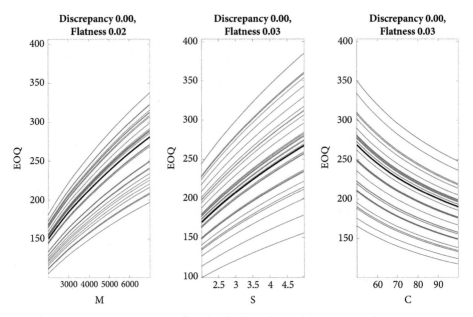

Fig. 5.5 Comparative statics with individual conditional expectation plots (Goldstein et al. 2015). The black line displays the average marginal behaviour of the EOQ as a function of each of the three inputs.

An insight which is also considered crucial for interpretability in machine learning is input importance. For analysts it is relevant to know the factors that drive a model response. Knowing, for example, that a given input is extremely relevant for the prediction of an algorithm would help analysts understand that the algorithm might be easily biased. If the input is discriminatory in nature, then the algorithm predictions themselves would become questionable from an ethical standpoint (see the discussion in Rudin 2019). Understanding feature importance goes under the setting of *factor prioritization* (Saltelli 2002, Saltelli et al. 2008). Analysts can find several methods that accomplish this in the literature. On the one hand, we have local methods such as elasticity or the differential important measure (Borgonovo and Apostolakis 2001), in which the indications of partial derivatives are synthesized in dimensionless indicators (indeed, partial derivatives themselves may not be directly comparable because they carry the units on which the input and output are denominated and inputs may have different units). On the other hand, in the well-known representation of tornado diagrams (Howard 1988), one plots the effects of one-at-a-time variations in the inputs. These methods are local. Alternatively, one can use global methods, within which there are regression-based methods (Kleijnen and Helton 1999), variance-based methods (Saltelli and Tarantola 2002), moment-independent methods (Borgonovo and Apostolakis 2001, Da Veiga 2015), and Shapley values (Owen 2014). The literature here is vast, and we refer the reader to works such as Razavi et al. (2021), Saltelli et al. (2008), Borgonovo (2017), and the handbook of Ghanem, Higdon, and Owhadi (2017) for further details. To illustrate the insights, let us present a graphical result obtained for our EOQ model.

The first triplet of bars in Fig. 5.6 represents the relative importance of the three inputs of our EOQ example, computed using variance-based importance measures. The second triplet displays the relative importance using a moment-independent sensitivity measure (the δ-importance—see Borgonovo and Apostolakis 2001). Both indices concur in evidencing M as the most important input, followed by S and C.[3]

The last setting we address is *model structure determination*. In this setting, the analyst wishes to understand whether the variations of the output are the sum of the variations induced by the individual inputs. If this is the case, then the model is said to be additive; conversely, interaction effects are important. The study of interactions is a huge area of modelling. Often, interactions are studied

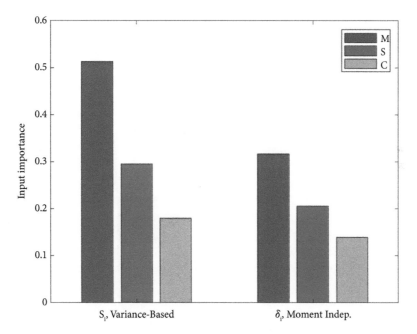

Fig. 5.6 Input importance measured with variance-based (left bars) and moment-independent sensitivity measures (right bars).

[3] For the technically inclined reader, we have computed these sensitivity measures from the same input–output dataset as that used for uncertainty quantification in Fig. 5.4. The technique is called given-data estimation and is documented in works such as Strong, Oakley, and Chilcott (2012) and Plischke, Borgonovo, and Smith (2013).

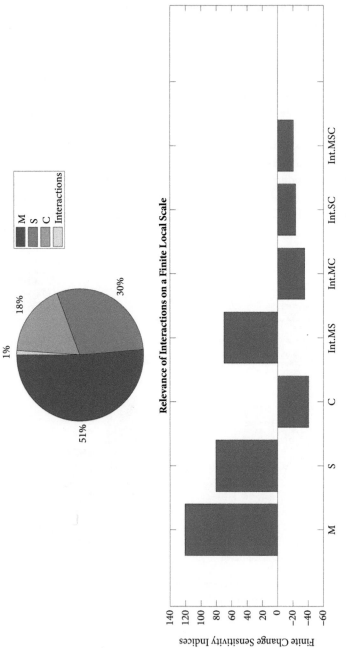

Fig. 5.7 Upper graph: Interactions on a global scale for the running example of our EOQ model. Lower graph: Interaction on a local scale for the same model.

in association with causality, as detailed in the book of Vanderweele (2015). Relevant to the determination of interactions is the scale at which the analyst is looking. We may consider interactions at a global, local, or even infinitesimal scale. To illustrate, let us carry out an analysis of interactions at the global scale for our problem. It can be seen that the difference between 1 and the sum of variance-based sensitivity measures is an indicator of global interactions (Owen 2003, Saltelli 1999). In our example, this difference is equal to 0.01, signalling that the model behaviour is additive over the range in question.

However, consider carrying out a local analysis of the response of the EOQ model when the inputs are between their extremes: we look at the difference between the EOQ computed when the inputs are fixed at [2000,2,50] and [700,5,100]. Such a difference is equal to 151.26. The literature has shown that this change can be exactly apportioned to the individual effects of M, S, and C, to their pairwise interactions, and to the remaining triple interaction. The result is displayed in the lower graph of Fig. 5.7. This graph shows that interactions matter in explaining the EOQ variation, when the inputs jump from the lower to the upper extreme of their ranges. In particular, we record positive (the interaction between M and S) as well as negative interactions (the ones between M and C, S and C, and M, S, and C). The main message here is that the scale at which we analyse the model structure—local or global—makes an important difference. Space limitations do not allow us to enter into the technical aspects of these results, but we would point the reader to the recent work of Borgonovo, Plischke, and Rabitti (2021).

Conclusion

We would like to close with some further reflections concerning the choice of the assumptions inspired by the critical review in Pfleiderer (2020). Hornberger and Spear (1981: 8) state: 'most simulation models will be complex with many parameters, state variables and non-linear relations. Under the best circumstances, such models have many degrees of freedom and, with judicious fiddling, can be made to produce virtually any desired behaviour, often with both plausible structure and parameter values.' In his essay, Pfleiderer (2020) observes that scientists might cherry pick assumptions to create models that produce a result they (consciously or unconsciously) aim at. Indeed, in principle, the right set of assumptions will lead to a model of the form that produces a predetermined outcome Pfleiderer (2020) actually illustrates several examples and proposes the use of filters on assumptions to prevent this from occurring. Pfleiderer (2020) also underlines the ethical implications of such choices. Similarly, it is possible to manipulate inputs so that a given model produces a desirable numerical value. In this chapter, we have made an introductory effort to illustrate that carefully dealing with assumptions and thoroughly performing sensitivity analysis should help analysts prevent this type of pitfall. The analysis must be as transparent as possible, even if it is at risk of displaying contradictory behaviour from the same model. However, this would allow stakeholders to make a fully informed decision about whether to retain or discard a model's results and predictions.

References

Apley, D.W. and J Zhu. 2020. 'Visualizing the Effects of Predictor Variables in Black Box Supervised Learning Models'. *Journal of the Royal Statistical Society. Series B: Statistical Methodology*, 82(4), 1059–86.

Apostolakis, G.E. 1990. 'The Concept of Probability in Safety Assessments of Technological Systems'. *Science*, 250(4986), 1359–64.

Begoli, E., T. Bhattacharya, and D. Kusnezov. 2019. 'The Need for Uncertainty Quantification in Machine Assisted Medical Decision Making'. *Nature Machine Intelligence*, 1(1), 20–3.

Borgonovo, E. 2017. *Sensitivity Analysis: An Introduction for the Management Scientist*, 1st edition, International Series in Management Science and Operations Research, Springer, New York.

Borgonovo, E. and G.E. Apostolakis. 2001. 'A New Importance Measure for Risk-Informed Decision Making'. *Reliability Engineering & System Safety*, 72(2), 193–212.

Borgonovo, E., M. Pangallo, J. Rivkin, L. Rizzo, and N. Siggelkow. 2022. 'Sensitivity Analysis in Agent-Based Modeling: A New Protocol'. *Computational Mathematics and Organization Theory*, 28, 52–94.

Borgonovo, E., E. Plischke, and G. Rabitti. 2021. 'Interactions and Computer Experiments'. *Scandinavian Journal of Statistics*, 49(3), 1274–303, doi: 10.1111/sjos.12560.

Borgonovo, E., M. Baucell, E. Plischke, J. Barr, and H. Rabitz. 2021. 'Trend Analysis in the Age of Machine Learning'. *SSRN*, 3, 1–24.

Cárdenas-Barrón, L.E., K.-J. Chung, and G. Treviño-Garza. 2014. 'Editorial: Celebrating a Century of the Economic Order Quantity Model in Honor of Ford Whitman Harris'. *International Journal of Production Economics*, 155, 1–7.

Da Veiga, S. 2015. 'Global Sensitivity Analysis with Dependence Measures'. *Journal of Statistical Computation and Simulation*, 85(7), 1283–305.

Frigg, R. and S. Hartmann. 2020. 'Models in Science'. In *The Stanford Encyclopedia of Philosophy*, edited by E.N. Zalta, Spring 2020 ed., Metaphysics Research Lab, Stanford University, Stanford, CA, pp. 1–10.

van der Gaag, L.C., S. Renooij, and V.M.H. Coupe. 2007. 'Sensitivity Analysis of Probabilistic Networks'. *Advances in Probabilistic Graphical Models, Studies in Fuzziness and Soft Computing*, 214, 103–24.

Ghanem, R., D. Higdon, and H. Owhadi, eds. 2017. *Handbook of Uncertainty Quantification*, Springer Verlag, New York.

Glasserman, P. 2003. *Monte Carlo Methods in Financial Engineering*, Springer Verlag, New York, NY.

Goldstein, A., A. Kapelner, J. Bleich, and E. Pitkin. 2015. 'Peeking Inside the Black Box: Visualizing Statistical Learning with Plots of Individual Conditional Expectation'. *Journal of Computational and Graphical Statistics*, 24(1), 44–65.

Griewank, A. 2000. *Evaluating Derivatives, Principles and Techniques of Algorithmic Differentiation, Frontiers in Appl. Math.*, vol. 19, SIAM, Philadelphia, PA.

Harris, F.W. 1913. 'How Many Parts to Make at Once'. *Factory, The Magazine of Management, 10(2)*, 135–6.

Hastie, T., R. Tibshirani, and J. Friedman. 2001. *The Elements of Statistical Learning,* vol. 1, Springer Series in Statistics, Springer, New York.

Hornberger, G.M. and R.C. Spear. 1981. 'An Approach to the Preliminary Analysis of Environmental Systems'. *Journal of Environmental Management, 12,* 7–18.

Howard, R.A. 1988. 'Decision Analysis: Practice and Promise'. *Management Science, 34(6),* 679–95.

Kleijnen, J.P.C. and J.C. Helton. 1999. 'Statistical Analyses of Scatterplots to Identify Important Factors in Large-Scale Simulations: 1: Review and Comparison of Techniques'. *Reliability Engineering & System Safety,* 65, 147–85.

Le Baron, B. 2006. 'Agent-Based Computational Finance'. *Handbook of Computational Economics, 2,* 1187–233.

Metropolis, N. 1987. 'The Beginning of the Monte Carlo Method'. *Los Alamos Science, 15,* 125–30.

Oakley, J.E. 2009. 'Decision-Theoretic Sensitivity Analysis for Complex Computer Models'. *Technometrics, 51(2),* 121–9.

Owen, A.B. 2003. 'The Dimension Distribution and Quadrature Test Functions'. *Statistica Sinica, 13,* 1–17.

Owen, A.B. 2014. 'Sobol' Indices and Shapley Values'. *SIAM/ASA Journal on Uncertainty Quantification, 2(1),* 245–51.

Pfleiderer, P. 2020. 'Chameleons: The Misuse of Theoretical Models in Finance and Economics'. *Economica, 87(345),* 81–107.

Plischke, E., E. Borgonovo, and C.L. Smith. 2013. 'Global Sensitivity Measures from Given Data'. *European Journal of Operational Research, 226(3),* 536–50.

Quarteroni, A., A. Manzoni, and C. Vergara. 2017. 'The Cardiovascular System: Mathematical Modelling, Numerical Algorithms and Clinical Applications'. *Acta Numerica, 26,* 365–90.

Rakovec, O., M.C. Hill, M.P. Clark, A.H. Weerts, A.J. Teuling, and R. Uijlenhoet. 2014. 'Distributed Evaluation of Local Sensitivity Analysis (DELSA), with Application to Hydrologic Models'. *Water Resources Research, 50(1),* 409–26.

Razavi, S., A. Jakeman, A. Saltelli, C. Prieur, B. Iooss, E. Borgonovo, E. Plischke, S. Lo Piano, T. Iwanaga, W. Becker, S. Tarantola, J.H.A. Guillaume, J. Jakeman, H. Gupta, N. Melillo, G. Rabitti, V. Chabridon, Q. Duan, X. Sun, S. Smith, R. Sheikholeslami, N. Hosseini, M. Asadzadeh, A. Puy, S. Kucherenko, and H.R. Maier. 2021. 'The Future of Sensitivity Analysis: An Essential Discipline for Systems Modeling and Policy Support'. *Environmental Modelling and Software, 137,* 104954.

Rudin, C. 2019. 'Stop Explaining Black Box Machine Learning Models for High Stakes Decisions and Use Interpretable Models Instead'. *Nature Machine Intelligence, 1(5),* 206–15.

Saltelli, A. 1999. 'Sensitivity Analysis. Could Better Methods Be Used?' *Journal of Geophysical Research, 104(D3)(37),* 89–93.

Saltelli, A. 2002. 'Sensitivity Analysis for Importance Assessment'. *Risk Analysis, 22(3),* 579–90.

Saltelli, A., G. Bammer, I. Bruno, E. Charters, M. Di Fiore, E. Didier, W.N. Espeland, J. Kay, S. Lo Piano, D. Mayo, R. Jr. Pielke, T. Portaluri, T.M. Porter, A. Puy, I. Rafols, J.R. Ravetz, E. Reinert, D. Sarewitz, P.B. Stark, A. Stirling, J.P. van der Sluijs, and P.

Vineis. 2020. 'Five Ways to Ensure That Models Serve Society: A Manifesto'. *Nature*, *582*, 482–4.

Saltelli, A., M. Ratto, T. Andres, F. Campolongo, J. Cariboni, D. Gatelli, M. Saisana, and S. Tarantola. 2008. *Global Sensitivity Analysis – The Primer*, Wiley, Chichester.

Saltelli, A. and S. Tarantola. 2002. 'On the Relative Importance of Input Factors in Mathematical Models: Safety Assessment for Nuclear Waste Disposal'. *Journal of the American Statistical Association*, *97*(*459*), 702–9.

Samuelson, P. 1947. *Foundations of Economic Analysis*, Harvard University Press, Cambridge, MA.

Strong, M., J.E. Oakley, and J. Chilcott. 2012. 'Managing Structural Uncertainty in Health Economic Decision Models: A Discrepancy Approach'. *Journal of the Royal Statistical Society, Series C*, *61*(*1*), 25–45.

Sullivan, T.J. 2015. *Introduction to Uncertainty Quantification*, Springer Verlag, Cham.

Vanderweele, T.J. 2015. *Explanation in Causal Inference*, Oxford University Press, New York, NY.

Wendell, R.E. 1985. 'The Tolerance Approach to Sensitivity Analysis in Linear Programming'. *Management Science*, *31*(*5*), 564–78.

Winsberg, E. 2019. 'Computer Simulations in Science'. In *The Stanford Encyclopedia of Philosophy*, edited by E.N. Zalta, Winter, 2nd edition, Metaphysics Research Lab, Stanford University, Stanford, CA.

6

Mind the consequences

Quantification in economic and public policy

Wolfgang Drechsler and Lukas Fuchs

Introduction

Governments use a wide array of quantitative methods, be it in the form of models, indicators, or statistics. Quantification permeates decision-making in economic policy; channels information about trade, inflation, opinions, unemployment rate, business confidence, economic growth, or economic inequality; and, most importantly, dominates the outward and inward communication about the aims, plans, and results of policy.

The ethical implications of decisions made by algorithms have received widespread attention in recent years. However, 'while algorithms appear to pose a clear and present danger, other instances of quantification—such as metrics, statistical inference and mathematical modelling, can also have an enduring and damaging influence' (Saltelli 2020: 2). Our focus is on quantification as such, and the use of these models and numbers in economic policy. A typical conceptual focus is what Muller calls 'metric fixation', which is 'the seemingly irresistible pressure to measure performance, to publicize it, and to reward it, often in the face of evidence that this just doesn't work very well' (2018: 4). This focus on metric fixation allows for at least some measurement: 'The problem is not measurement, but excessive measurement and inappropriate measurement—not metrics, but metric fixation' (ibid.). As we will see, however, this is too fast and too easy.

Of course, there are strong arguments in favour of the use of quantifiable indicators in policy. Already, the focus on metric fixation implies that there may be beneficial uses of numbers in government.[1] Metric fixation is intertwined with widely accepted norms of objectivity and how economics can play a justificatory role in public (see 'Modelling as closure for justification' in chapter 7). Fixing our practice will not suffice. Instead, we need to conceive a new paradigm in economic thought and practice.

---

[1] For further discussion of negative consequences, see Muller (2018: chapter 2), Talbot (2005) and Power (2005); Drechsler (2000); Drechsler (2019); also Drechsler (2020).

Wolfgang Drechsler and Lukas Fuchs, *Mind the consequences*. In: *The politics of modelling*. Edited by: Andrea Saltelli and Monica Di Fiore, Oxford University Press. © Oxford University Press (2023). DOI: 10.1093/oso/9780198872412.003.0006

This paradigm, we suggest, could be informed by intellectual traditions that provide the conceptual tools for nuanced understanding in economics, namely hermeneutics and the German Historical School of Economics (GHS). One of us (Wolfgang Drechsler) has drawn heavily on these two traditions in his discussions on 'whether quantitative-mathematical social science is at all possible' (2000: 246; see also 2011). The conclusion of these papers is negative: numbers misrepresent social phenomena, and this applies equally to the realm of economics. The present chapter will follow the ideas of these intellectual traditions to better understand the negative consequences of overreliance on numbers, and models, in economic policy.

The first negative consequence of metric fixation is epistemic. Overreliance on numbers results in limited understanding of the phenomena and policy options under consideration. Such a shallow and one-sided understanding is subject to distortions and serious flaws. We contend that the resulting epistemic limitations cannot be overcome through better maths or more sophisticated modelling.

The numbers critique

A plain number can be illusively simple yet convincing. The *Manifesto* makes this observation: 'Once a number takes centre-stage with a crisp narrative, other possible explanations and estimates can disappear from view' (Saltelli et al. 2020: 484). It can dramatically narrow the depth of our comprehension and the range of policy alternatives. Therefore, there are serious problems with such a distorted view on reality. In economic policy especially, the focus on simple yet illusive numbers has limited our understanding and the range of policy choices. The notion that economic policy must chiefly be conducted and communicated through numbers is pervasive.

The limits of numbers in depicting ideas have received renewed attention in recent years: in particular, their inability to express genuine uncertainties (King and Kay 2020) or their susceptibility to distortion through gaming, lowering standards, omission, or cheating (Muller 2018: 23–5). We may accept this challenge and devise new indicators and more sophisticated mathematical methods that can overcome these limitations. However, we disagree that the problem here is merely 'bad maths'—that is, the limitations of our current methods and indicators. We press the use of numbers on more general grounds: 'the things themselves', the proper objects of enquiry in the social sciences, are not quantifiable without causing serious damage to our understanding. Even if these technical problems of expressing concepts in numbers were resolved, and even if we could build systems of data collection that could not be subverted in the ways just listed, there would still be serious limitations to understanding based mostly on numerical metrics.[2] No amount of technical sophistication and tinkering with definitions will make quantification a fully adequate approach. Instead, we should aim for a different standard of understanding in the social sciences.

Heinrich von Stackelberg played a key role in introducing mathematical methods into German Economics in the middle of the twentieth century. In his *Grundlagen der theoretischen Volkswirtschaftslehre*, he defends the use of mathematics by considering an objection, namely that:

> mathematics would fake an exactness and rigidity of economic relations which in reality would be flowing and inexact; it would fake necessities of natural science laws which in reality the human would be able to decide and shape freely.

Von Stackelberg's response is instructive:

> This view completely mistakes the role of mathematics in economic theory. How often has it been said from the expert side that 'there never jumps more out of the mathematical pot than has been put into it before'! Mathematical symbolics changes neither the preconditions nor the results of the theoretician, as long as they are concludent.
>
> (von Stackelberg 1951, x–xi)

The defence is, then, that mathematical formulation is harmless. Its role consists merely in making ideas clear and precise. No substantial theoretical assumptions are added into the analysis by including these techniques. However, this defence is fragile, for the following reasons.

For numbers to express quantities, the objects falling under a concept must be made identical in this respect. This mode of numbers—commensuration—'creates a specific type of relationship among objects. It transforms all difference into quantity' (Espeland and Stevens 2008: 408). Similarly, Ravetz (1971/1996: xv) argues that 'the process of standardization necessarily suppresses anything that might be confusing; and so the presence of obscurities at the foundations of theoretical science is a taboo subject, among teachers and publicists'. This process of standardization blends out much information, some of which may be redundant, but some of which may be crucial. How recent and contingent this process of standardization in economic phenomena has been is illustrated by the case of revolutionary France. In the process of producing a uniform survey, it was found that France

---

[2] There are at least two more ways in which numbers could be subverted. First, numbers may be used to display logic when reasoning from outlandish assumptions, called *folie raisonnante* in French (Porter 2012b: 587). Second, statistical (e.g. p-hacking) and modelling malpractices (e.g. adjusting models to clients' wishes) are another way in which otherwise good numbers can be subverted to arrive at epistemologically dubious conclusions (see Saltelli 2020: 5).

'was still too divided by status and locale to fit into the categories prescribed by the statisticians' (Espeland and Stevens 2008: 411). A more centralized state and uniform population was required in order to produce the homogeneity and technical skills required to produce these kinds of numbers.

The resulting numbers only provide a shallow picture of the underlying phenomena. Porter (2012a) speaks of 'thin description' in this context.[3] It is *thin* because it provides a lean and standardized representation of a phenomenon, a representation that can be easily comprehended without the need for in-depth understanding of the meaning, context, and embeddedness of the phenomenon in question. In the economic realm, the success of a company or policy can be judged by some simple numerical measures, such as profit or growth. For Porter, the usage of such numbers came to prominence due to 'an ethic of thin prescription' (2012b: 595) that favours judging 'a person or institution by a few numbers or, ideally, one number'. This is the world of the single unit ranking—so common in comparisons between applicants, economies, universities—that gives off the appearance of objectivity. In economic policy, the increasing reliance on cost–benefit and risk analysis demonstrate that increasing ethic of thin description (they become 'numbers of neoliberalism' (Porter 2012b: 593)).

This connection between standardization and loss of quality of information has a further dimension: the resulting information presents itself as objective and authoritative. Caveats, inherent unclarities, and uncertainties are ignored, and 'nothing does more to create the appearance of certain knowledge than expressing it in numerical form' (Muller 2018: 24; see Gupta 2001). As a result, the numbers give off the impression of an exhaustive representation of the phenomenon in question. Jasanoff's concept of *technologies of hubris* is useful here. These technologies generate overconfidence in their results due to their appearance of exhaustivity (see also 'Explosion of the uncertainty space' in chapter 4). These methods 'focus on the known at the expense of the unknown' (Jasanoff 2005: 238; see also 'The concealed politics of hidden unknowns' in chapter 7).

The emphasis on numbers in economic policy is closely linked with economic science, which may be one such technology of hubris. Mainstream neo-classical economics, taught in the form of textbooks in nearly all economics courses in the world and dominating prestigious economics journals—'Standard Textbook Economics' (STE)—depends heavily on modelling and formalization. These techniques give it the mantle of epistemic credibility, to the point that if it cannot be modelled or formalized, it is non-scientific and not even economics. While this might be a peculiar standard within a profession, it has also had consequences for the social boundaries of this profession. The field became closed off from other disciplines: 'The technical training of modern economists leaves them more able than

---

[3] The concept of 'thick description', against which 'thin description' is contrasted, is from Geertz (1973).

their teachers to talk to mathematicians, statisticians, and engineers, but less able to talk to interpretive historians, sociologists, anthropologists, or philosophers' (Lavoie 1992: 3). As a result, the interpretation of economic phenomena in the public has become reserved to economists and mathematicians.

As a first step, these experiences with the overreliance on numbers might warrant a call for moderation. Excesses of metric fixation should be avoided in order to find a golden middle path between overreliance and not using numbers at all. However, keeping down the excess is only one way to address this problem. A further and more fundamental way to address this gap of understanding in economics would be to consider a more substantial paradigm shift in how we apply epistemic concepts to the economy. Such an alternative paradigm might take inspiration from hermeneutics.

The Hermeneutic approach

Hermeneutics is a philosophical tradition that offers a unique and attractive alternative to the epistemology underlying a metric-fixated method. This philosophical tradition puts emphasis on holistic, context-sensitive, and meaning-oriented understanding (*Verstehen*). The central reference point for this philosophical tradition is the work of the late Hans-Georg Gadamer, in particular his *Truth and Method* (1960/1990).[4] One key insight of this book is that one could form a juxtaposition—truth *or* method, method *or* truth—as opposed to the assumption that method actually produces truth, on which 'empirical' social science is often based. Gadamerian hermeneutics has had an enormous impact on all humanities and some social sciences. In contemporary philosophy, hermeneutics and its larger context of 'Continental Philosophy' is probably the most orthodox heterodoxy today, even within Anglo-American academic institutions.

One of Gadamer's key concerns was to point out the fundamental difference between natural and social science: 'The experience of the social-historical world cannot be lifted up to science by the inductive process of the natural sciences' (Gadamer 1960/1990: 10). Natural sciences deal with objects, social ones with subjects, i.e. with human beings. We can trace the implications of this hermeneutical insight in economics: 'this basic difference has a decisive impact on the transferability of concepts from one to the other' (Drechsler 2011: 47).

Economic policy and economic science following such an alternative epistemic paradigm might have the advantage of displaying its value judgements, assumptions, and cultural context more openly. It would constitute a turn away from a

---

[4] Lawn (2006) is perhaps the best introduction to Gadamerian hermeneutics available in English. See Drechsler (2016b) for one of the author's personal takes and Drechsler (2004) for the explicit connection between hermeneutics and economics.

view of numbers as a central and authoritative means to truth, towards a more modest incorporation of numbers into a holistic method that recognizes economic phenomena as complex problems.

Economics is a social science, and the consequences of hermeneutical thinking on the social sciences have received the most attention, especially in Gadamer's work. For this reason, we start by tracing the hermeneutical critique in the social sciences before we apply them more specifically to the case of economics in the next section.

The success of natural sciences in the past centuries has made social scientists look to the natural sciences for a 'rigorous' methodology, a phenomenon that has been dubbed 'physics-envy'. If one should conceive of the social sciences as something somehow 'between' the natural sciences and the humanities, then the quantitative-mathematical kind has a very strong tendency towards the natural sciences side. The social sciences have taken the natural sciences as their role model. As a result, we find strong tendencies towards positivist thinking in the social sciences. The increased reliance on quantification and formal methods reflects this shift in the research methodology. Wilhelm Lexis, an important scholar in the GHS, spelled this out as follows:

> Right away, a certain analogy is noticeable which exists between the social and the natural sciences. The means of realization for the one as well as for the other is supposed to be experience. As the natural sciences are taken to be the specific empirical sciences, the temptation is close at hand to put the social sciences under the guidance of her older sister by directing to her the tried method of the latter.
>
> (Lexis 1903: 235)

Where does this emulation of the methodology leave the social sciences? We must ask of a theory that it mirrors reality, that it delivers meaningful statements about reality. Of course, reality is muddy and messy. If a theory does not mirror reality, however, it is not true and is therefore wrong. There are strong reasons to think that the research methodology of the natural sciences leaves the social sciences unable to mirror reality. The tools on which it is based misrepresent human interaction and thus human life, which is the object of social sciences. The key issue with quantification and positivistic methodology is that it is unable to distinguish between those phenomena that can be measured and those that cannot.

The importance of Gadamerian hermeneutics for the social sciences is implied in G.B. Madison's reminder that 'the universality of Hermeneutics is based solely on the hermeneutical *fact* that [...] what makes human beings "human" is their "linguisticality"' (Madison 1997: 360). To understand is not less, or less scientific, than to assess from the outside, as in the natural science world; but it is more, or more so. Werner Sombart has put this very well, in terms quite similar to the thesis of *Wahrheit und Methode*: The natural sciences' successful attempt to monopolize the truth is a reversal of the real situation: "'True' realisation reaches as far as we "understand", that is, it is limited to the area of culture and fails towards nature' (Sombart 1956: 9). As Nicolai Hartmann put it, 'Only "meaning" can be "understood" as well as all that which is related to it: value, goal, significance' (1951: 33; 64–76). And Sombart: 'Realisation that wants to arrive at the being of nature is metaphysics' (1956: 75). This means that although we cannot talk very meaningfully about things in biology and physics, the situation in 'the social sciences is completely different: here, our realisation is capable of immediate penetration of the inner causal connection of the outer phenomena' (Lexis 1903: 242–3). It does not mean, of course, that understanding inevitably leads to the truth, but it means that there is a chance that it does, or might.

While the arguments from the last paragraphs all apply to economics *qua* social science, another concern has particular relevance for this field. The objects studied by economics, the people, goods, institutions, production or exchange, have particular relevance for our practical life. Economics is about the economy; it is not merely an intellectual playground but matters greatly for human life, individual happiness, and social justice. It is our task in economics and the social sciences to consider whether a calculated world is a better one, difficult and tedious as this deliberation may be, particularly if one wants to draw conclusions and maybe even implement them. The question is what life we as human beings, in the context of structured living-together in time and space, want to live, and whether quantitative-mathematical thinking and its claims to an adequate representation of the world actually make a better or a worse world.

> Only if science is subordinated to *phronēsis* could it fulfil the task of the future. [...] Wherever methods are being employed their correct application is not specified by a method but demands our own judgement. This is a profound commonality of reason itself. It testifies to the depth in which linguisticality is rooted in human life. All methods require judgement and linguistic instruction.
>
> (Gadamer 1997: 366–7)

How can we go on, however, in concrete economic research terms, especially in light of the practical tasks economics has to fulfil? This question leads us to a look at normativity. Its re-inclusion in the social sciences would be a return to the Greeks, at least in perspective—and specifically to the Greeks, and especially Aristotle, as seen by the GHS and by hermeneutics. And here, understanding and normativity are linked in a way as to produce a possible, meaningful, truth-focused approach: 'The Aristotelian program of a practical science seems [...] to be the only science-theoretical model according to which the "understanding" sciences can be thought' (Gadamer 1997: 87).

The Historical School

The GHS is a tradition that approaches questions of the measurement of economic phenomena and the respective contributions of market and state in a modest way. This school was part of a set of historical schools at the time (such as the Historical School of Jurisprudence). They were 'historical' in the sense that they focused on the specificities of historical developments, as opposed to seeking perennial, context-free truths. This school dominated economics in Germany at the turn of the twentieth century, both academically and in policy. The GHS differs from the current economic paradigm in key aspects (Seligman 1925: 15–16). First, it discards the deductive method, preferring historical context and use of statistics (see also Porter (1995: 54)); second, it rejects the assumption that there are immutable laws in economics and instead appreciates the embeddedness of theories and social institutions; and third, it adopts a more nuanced attitude towards the role of the state in the economy, which has found its practical consequences in the social policy advocated by many of these scholars.[5]

In comparison to hermeneutics, the GHS is less known in academic discourse today and remains obscure, notwithstanding several attempts at re-evaluation and resurrection.[6] Even current initiatives to give heterodox economics a greater place on the stage—often from an Anglo-American perspective, even if this is rejected—generally neglect the GHS. However, this school shares important tenets of the hermeneutical tradition, especially the 'younger' (headed by Schmoller) and 'youngest' (headed by Sombart) generations. The GHS dominated economics in Germany and well beyond for the last few decades of the nineteenth century and the first one or two of the twentieth, both academically and, since the 1890s, in policy.

For Sombart (1956: 78), 'this insight is of extraordinary significance: holistic understanding—the understanding of structure—is what distinguishes the humanities'. Key to this is Sombart's premise that social sciences can aim at *Sinneszusammenhänge*, the larger context or significance of social phenomena. Only understanding of this larger context can allow us to fully understand particular phenomena. The fixation on metrics and other quantified methods may not only distract us from but even undermine such a holistic understanding of the broader economic world we live in.

The role of institutions and culture in explaining individual economic facts was highlighted by Schmoller. In 1897, he served as rector of the University of Berlin, then arguably the best university in existence. In his inaugural address, he narrated

---

[5] Probably the most consequential achievement of this school was the development of the welfare state in Bismarck's German Empire at the end of the nineteenth century, which was the forerunner to the modern welfare state generally (Drechsler 2016a).

[6] We only mention, for their work in English, Jürgen G. Backhaus, Ha-Joon Chang, Geoffrey Hodgson, and Erik S. Reinert.

the story of how the GHS has become standard (and textbook) economics, conscious of his school's accomplishments, but in hindsight, we know that these words were almost an epitaph. Nonetheless:

> Thus, a mere science of market and exchange, a sort of business economics which threatened to become a class weapon of the property owners, returned to being a great moral-political science which examines both the production and the distribution of goods, as well as both value and economic institutions, and which has its central focus not on the world of goods and capital but on the human person.
>
> (1904: 388)

It is important to note that the GHS was not opposed to the use of statistics in its methodology. Porter points this out:

> It is curious but revealing that, in the *Methodenstreit* between the historicist followers of Gustav Schmoller and the deductivist Austrian school of Carl Menger, quantification was plainly on the side of history. Though antideductive, it provided from another standpoint a middle way between the verbal theories of Menger and the new mathematical marginalist theory that Lexis criticized as excessively abstract. Deductive theory, he charged, can show no more than tendencies. Its propositions do not give a 'reliable predetermination of actual events, and cannot by themselves decide the measures to be taken in pursuit of goals in economics'. For the historical school, the goals of economics were first of all practical and administrative ones. Its members aimed above all at social reforms, to improve the lives of workers. Effective state intervention in economic affairs, they believed, depended on expertise that had proved itself by its empirical adequacy. This of course was more easily said than achieved. But given the choice, they preferred descriptive accounts and statistics to formal, deductive theory. The same outlook was typical of most natural scientists who wrote on economic matters.
>
> (Porter 1995: 54)

The GHS may thus offer some inspiration for a future economic paradigm in which quantification is not the central measure for scientific relevance, but—if one may use this loaded word—reality is.

The public sector

Having considered in some depth how metric fixation distorts our understanding of economic phenomena, we can now also discuss two more negative consequences in the GHS context. The second consequence of overreliance on numbers in economic policy concerns the relationship between citizens and public officials. When the actions of politicians and civil servants are subject to scrutiny, numbers appear to offer a solid basis for them to legitimise their decisions. The classic recent study of this role of numbers in public discourse is, again, Porter (1995), but for the GHS, and especially Schmoller, public discourse, public sector performance, and the role of the state were key issues already.

For Porter, in a complex society, civil servants are faced with a great number of decisions, and it is impossible to get a popular mandate for all decisions. Public officials face the question of how this discretion can be exercised without being accused of arbitrary or even irresponsible action: 'faith in objectivity tends to be associated with political democracy, or at least with systems in which bureaucratic actors are highly vulnerable to outsiders' (Porter 1995: 8). This problem is exacerbated not only by the plurality of conceptions of the good in society, but also by epistemic disagreement, such as divergent views on what constitutes solid justificatory evidence for the likely success of a policy. In the face of such unstable grounds on which public decisions can be justified, the reference to numbers appears as objective, unambiguous, and authoritative. Porter (1995: 8) expresses this point succinctly: 'The appeal of numbers is especially compelling to bureaucratic officials who lack the mandate of a popular election, or divine right [...] Quantification is a way of making decisions without seeming to decide.' Espeland and Stevens (2008: 417) list four ways in which numbers may display authority: 1) through their sense of accuracy or validity; 2) through their usefulness in solving problems; 3) by linking users who have investments in numbers; and 4) through their association with rationality and objectivity.

In the public sector, the fixation on performance indicators has implications for public management reform. New Public Management (NPM), the paradigm in public administration that orients itself towards business practices, places great emphasis on performance measurements and their use in decisions, promotions, and policy. However, such a close connection between numbers and legitimacy results, *inter alia*, in civil service timidity. Unless they have the numbers that are perceived to provide the necessary legitimacy, they are unable to carry out more ambitious policies. This restriction also results in unambitious public agents who have to rely on external producers of legitimacy (consultancy firms).[7] What started off as a plausible democratic ideal—that civil servants should act on objective standards and be held accountable—may easily be, and has generally been, subverted by government inaction and timidity.[8]

---

[7] Aside from consultants, external legitimacy can also be sought from well-endowed (in data and methodology) private actors in a process that opens the door to regulatory capture; see Drutman (2015); Laurens (2017); Saltelli et al (2022).

[8] NPM fully embraces the 'quantification myth', namely that 'everything relevant can be quantified' (Drechsler 2005: 1). On the need for bold and imaginative government, see Mazzucato (2021).

The ideology of numbers

The third and final negative consequence concerns hidden normative economic assumptions. The fixation on quantification means that those things that produce measurable benefits are valued more highly than those whose benefits cannot be measured. The converse remark applies to costs. In this way, the overreliance on numbers in scientific and public discourse may promote biased normative economic views.

In particular, the presumption in favour of unrestricted markets in political economy is closely related to the need to justify economic arrangement by reference to numbers. Of course, the fixation on metrics and numbers is not the only basis; it forms a symbiotic relationship with the assumptions and practices in neo-classical economics in STE. The combination of these—the authority of numbers and the simplicity of the assumptions in neo-classical economics—makes these a formidable obstacle to gaining a sociological, nuanced, and context-sensitive understanding of economics and policy.

Numbers have an important relationship to the laissez-faire economic paradigm. They 'help constitute the things they measure by directing attention, persuading, and creating new categories for apprehending the world' (Espeland and Stevens 2008: 412). This 'reactive' nature of numbers over time ossifies our descriptive and normative views on the economy by partly constituting the facts to be observed.[9]

The impressive array of theorems and mathematical models that are adduced by defenders of the maths-based defence of economic regimes put the critic in the defensive position. In order to reject the defence, one would have to express the challenge in terms of the dominant discourse, otherwise it would be prone to the criticism that one did not work through the relevant models. In this way, these metrics and models may 'pre-empt political discussion', another factor that Jasanoff identifies in *technologies of hubris*. While these models appear to be objective and neutral, they hide and perpetuate important exercises of judgements— sometimes technical, sometimes normative. Consequently, opportunities for challenging the models are reserved for those who master the technology and who are able to put the challenge in terms of the analysis that is under scrutiny (Jasanoff 2005).

As we saw already, these problems of quantification and indicators are especially pertinent in and for government. Henry Mintzberg (1996: 79) wrote:

Things have to be measured, to be sure, especially costs. But how many of the real benefits of government activities lend themselves to such measurement? [...]

---

[9] Other examples for this 'reactive' or 'constitutive' role of numbers are the use of university rankings (Muller 2018: chapter 8; Espeland and Stevens 2008: 416) and the role of metrics for the recognition and acceptance of the LGBT+ movement (ibid.: 413–414). See also the discussion about the 'performativity' of economics in MacKenzie, Muniesa, and Siu (2007).

Many activities are in the public sector precisely because of measurement problems: If everything was so crystal clear and every benefit so easily attributable, those activities would have been in the private sector long ago.

And yet there is great emphasis on exactly that as exemplified by this quote by Angus Deaton, who won the 2015 Nobel Prize in Economics: "Progress cannot be coherently discussed without definitions and supporting evidence. Indeed, an enlightened government is impossible without the collection of data" (2013: 15)

In particular, the notion of Gross Domestic Product (GDP) as a measure of the economic wealth of a society has dominated policy discourse but has also received its share of criticism. Recently, Mazzucato (2018: chapter 8) argued that because of the focus on GDP, the contribution that government makes to society has been systematically underestimated. The problem with calculating the value produced by government is that its contributions are usually not offered in the market sphere where it would fetch a price. However, the price fetched is exactly the basis for value in the calculation of GDP (to the extent that it aims at measuring value at all). For example, when government provides security for its population or when it runs public utilities companies, it is unreasonable to take the price that these goods and services fetch as the basis for calculating its value. It seems that the GDP framework imposes an alien system of values on government. Of course, GDP is only one such metric; yet this illustrates how a seemingly objective measure can contain normative biases that have a profound impact on discourse and policy.[10]

Outlook Express

As a desirable consequence of our deliberations, then, a first ameliorative step would be to restrict the force of numerical concepts in shaping economic policy through a clearer understanding of the limits of these concepts and more responsible use. However, we again contend that—more ambitiously—we need to go beyond repairing a broken paradigm towards fresh thinking inspired by alternative schools of economic thought.

Charles Taylor wrote 'we could gain a great deal by examining our theorizing about social matters as a practice' (1983: 91). The use of indicators in government, too, is one social practice, and a very strong and firm one, but not a necessity. To use indicators and, in fact, to quantify, is not natural, but a choice—on the governance level as anywhere else. And every social practice comes with costs and with inherent limitations, given its aims.

---

[10] The Bhutanese alternative to GDP, 'Gross National Happiness' (GNH), was originally anti-quantitative, but morphed into a heavily indicator-driven concept once it became globally noticed and attractive (Drechsler 2019). GNH is, therefore, an example of both how a policy-relevant alternative to GDP is possible, and how pervasive the numbers-models ideology that GDP signifies actually is.

The most important limitation of this practice—the use of numbers in government—is that it diverts attention away from what matters and towards what is measurable. As we argued, the use of numbers may result in a separation of worlds, where everything must be in the books, and if it is not, it does not exist. In opposition to von Stackelberg, who claimed that maths does not do any harm, we argued that some of the most important phenomena in the social sciences cannot be comprehended through numbers. Policy that aims to understand problems and to make decisions mainly through numbers must therefore be inherently weak.

To conclude, the fixation on numbers in economic and public policy has negative consequences for our understanding of economic life and of the polities in which we live generally; it rightly undermines trust and accountability in public discourse and precludes the discussion of normative views that should be subject to that very discourse. An alternative way of attending to economic and generally public life and action—one that focuses on context-sensitive and holistic understanding—may benefit from the use of numbers and even sometimes models, but it cannot be driven by them.

References

Deaton, A. 2013. *The Great Escape: Health, Wealth and the Origins of Inequality*, Princeton University Press, Princeton, NJ.

Drechsler, W. 2000. 'On the Possibility of Quantitative-Mathematical Social Science, Chiefly Economics'. *Journal of Economic Studies*, 27(4–5), 246–59.

Drechsler, W. 2004. 'Natural vs Social Sciences: On Understanding in Economics'. In *Globalization, Economic Development, and Inequality: An Alternative Perspective*, edited by Erik S. Reinert, Edward Elgar, Cheltenham, pp. 71–87.

Drechsler, W. 2005. 'The Rise and Demise of the New Public Management'. *Post-Autistic Economics Review*, 33(14), 17–28.

Drechsler, W. 2011. 'Understanding the Problems of Mathematical Economics'. *Real-World Economics Review*, 56, 45–57.

Drechsler, W. 2016a. 'Kathedersozialismus and the German Historical School'. In *Handbook of Alternative Theories of Economic Development*, Edward Elgar, Cheltenham, pp. 109–23.

Drechsler, W. 2016b. *Gadamer in Marburg*, 3rd edition, Blaues Schloss, Marburg.

Drechsler, W. 2019. 'Kings and Indicators: Options for Governing without Numbers'. In *Science, Numbers and Politics*, edited by M.J. Prutsch, Palgrave Macmillan, Cham, pp. 227–62.

Drechsler, W. 2020. 'Seamlessness as Disenfranchisement: The Digital State of Pigs and How to Resist It', *Acta Baltica Historiae et Philosophiae Scientiarum*, 8(2), 38–53.

Drutman, L. 2015. *The Business of America Is Lobbying: How Corporations Became Politicized and Politics Became More Corporate*, Oxford University Press, Oxford.

Espeland, W.N. and M.L. Stevens. 2008. 'A Sociology of Quantification'. *Archives Européennes de Sociologie/European Journal of Sociology/Europäisches Archiv für Soziologie*, 49(3), 401–36.

Gadamer, H.-G. 1960 (1990). *Wahrheit und Methode. Grundzüge einer philosophischen Hermeneutik*, Mohr/Siebeck, Tübingen.

Gadamer, H.-G. 1997. 'Reply to G.B. Madison'. In *The Philosophy of Hans-Georg Gadamer*, The Library of Living Philosophers, Vol. XXIV, edited by L.E. Hahn, Open Court, Chicago and La Salle, IL, pp. 366–7.

Geertz, C. 1973. *The Interpretation of Cultures*, Basic Books, New York.

Gupta, S. 2001. 'Avoiding Ambiguity: Scientist Sometimes Use Mathematics to Give the Illusion of Certainty'. *Nature*, *412(6847)*, 589.

Hartmann, N. 1951. *Teleologisches Denken*, 2nd edition published in 1966, De Gruyter, Berlin.

Jasanoff, S. 2005. 'Technologies of Humility: Citizen Participation in Governing Science'. In *Wozu Experten?* edited by A. Bogner and H. Torgersen, VS Verlag für Sozialwissenschaften, Wiesbaden, pp. 370–89.

King, M. and J. Kay. 2020. *Radical Uncertainty: Decision-Making for an Unknowable Future*, Hachette UK, London.

Laurens, S. 2017. *Lobbyists and Bureaucrats in Brussels: Capitalism's Brokers*, Routledge, Abingdon.

Lavoie, D., ed. 1992. *Economics and Hermeneutics*, Routledge, Abingdon.

Lawn, C. 2006. *Gadamer: A Guide for the Perplexed*, Continuum, London.

Lexis, W. 1903/1874. 'Naturwissenschaft und Sozialwissenschaft'. In *Abhandlungen zur Theorie der Bevölkerungs- und Moralstatistik*, Gustav Fischer, Jena.

MacKenzie, D.A., F. Muniesa, and L. Siu, eds. 2007. *Do Economists Make Markets? On the Performativity of Economics*, Princeton University Press, Princeton, NJ.

Madison, G.B. 1997. 'Hermeneutics' Claim to Universality'. In *The Philosophy of Hans-Georg Gadamer*, The Library of Living Philosophers, Vol. XXIV, edited by L.E. Hahn, Open Court, Chicago and La Salle, IL, pp. 350–65.

Mazzucato, M. 2018. *The Value of Everything: Making and Taking in the Global Economy*, Hachette UK, London.

Mazzucato, M. 2021. *Mission Economy: A Moonshot Guide to Changing Capitalism*, Allen Lane, London.

Mintzberg, H. 1996. 'Managing Government, Governing Management'. *Harvard Business Review*, May/June, 75–83.

Muller, J.Z. 2018. *The Tyranny of Metrics*, Princeton University Press, Princeton, NJ.

Porter, T.M. 1995. *Trust in Numbers: The Pursuit of Objectivity in Science and Public Life*, Princeton University Press, Princeton, NJ.

Porter, T.M. 2012a. 'Thin Description: Surface and Depth in Science and Science Studies'. *Osiris*, *27(1)*, 209–26.

Porter, T.M. 2012b. 'Funny Numbers'. *Culture Unbound*, *4(4)*, 585–98.

Power, M. 2005. 'The Theory of the Audit Explosion'. In *The Oxford Handbook of Public Management*, edited by E. Ferlie, L.E. Lynn, and C. Pollitt, Oxford University Press, Oxford, pp. 326–44.

Ravetz, J. 1971 (1996). *Scientific Knowledge and Its Social Problems*, Clarendon, Oxford.

Saltelli, A. 2020. 'Ethics of Quantification or Quantification of Ethics?' *Futures*, *116*, 1–11.

Saltelli, A., G. Bammer, I. Bruno, E. Charters, M. Di Fiore, E. Didier, W.N. Espeland, J. Kay, S. Lo Piano, D. Mayo, R. Jr Pielke, T. Portaluri, T.M. Porter, A. Puy, I. Rafols, J.R. Ravetz, E. Reinert, D. Sarewitz, P.B. Stark, A. Stirling, J.P. van der Sluijs, and P.

Vineis. 2020. 'Five Ways to Ensure That Models Serve Society: A Manifesto'. *Nature*, *582*(*7813*), 482–4.

Saltelli, A., D.J. Dankel, M. Di Fiore, N. Holland, and M. Pigeon. 2022. 'Science, the Endless Frontier of Regulatory Capture'. *Futures*, *135*, 102860.

Schmoller, G. 1897/1904. 'Wechselnde Theorien und feststehende Wahrheiten im Gebiete der Staats- und Sozialwissenschaften und die heutige deutsche Volkswirtschaftslehre'. In *Über einige Grundfragen der Socialpolitik und der Volkswirtschaftslehre*, 2nd edition, Duncker and Humblot, Leipzig, 365–93.

Seligman, E.A. 1925. *Essays in Economics*, Macmillan, New York.

Sombart, W. 1956. *Noo-Soziologie*. Duncker & Humblot, Berlin.

von Stackelberg, H. 1951. 'Vorwort zur ersten Auflage'. In *Grundlagen der theoretischen Volkswirtschaftslehre*, 2nd edition, Polygraphischer Verlag, Tübingen, Zürich, Mohr/Siebeck, pp. viii–x.

Talbot, C. 2005. 'Performance Management'. In *The Oxford Handbook of Public Management*, edited by E. Ferlie, L.E. Lynn, and C. Pollitt, Oxford University Press, Oxford, pp. 491–518.

Taylor, C. 1983. *Philosophical Papers: Volume 2, Philosophy and the Human Sciences*, Cambridge University Press, Cambridge.

7

Mind the unknowns

Exploring the politics of ignorance in mathematical models

Andy Stirling

Introduction: The concealed politics of hidden unknowns

In both science and policy, a 'model' is the name routinely given to a simplified
and explicitly codified representation of what (under a particular view, in a specific moment, for a given purpose), amounts to relevant knowledge about some
bounded aspect of the world.

As such (no matter how sophisticated they may be), all models will not just be
radically insufficient with respect to the real world *as it actually is* (Saltelli et al.
2020a). They will also typically be seriously incomplete about the world as it is
variously *known to be*—reducing complexities under the particular represented
view(s), as well as entirely missing alternative understandings under perspectives
that are not included (Schwarz and Thompson 1990). Perhaps most momentously
absent in models are those (often many) aspects about which even the most expert
understandings are ignorant (Loasby 1976).

These are three crucial ways in which what remains 'unknown' around any
given model can often be more salient than the knowledge that is encoded.
Models sideline all-important uncertainties (Thompson and Warburton 1985),
gloss major ambiguities (Stirling 2003), and exclude crucial ignorance (Wynne
1992). With 'devils in details' in modelling even more than many other fields, the
implications can be profound. Although invoking the supposedly transcendent
authority of mathematics, modelling results depend strongly on: limits to prevailing understandings; the particular questions asked; assumptions that are made;
data that is available (or not); values that are prioritized; taken-for-granted judgements; presumptions that the past is a reliable guide to the future; the manner
in which findings are expressed; and the cultures and political arenas in which
they are interpreted (Saltelli et al. 2020a). Typically seriously neglected in policy
debates around particular models, many of these key general features of modelling
themselves often count among the most important unknowns (Brooks 1986).

Andy Stirling, *Mind the unknowns*. In: *The politics of modelling*. Edited by: Andrea Saltelli and Monica Di Fiore, Oxford
University Press. © Oxford University Press (2023). DOI: 10.1093/oso/9780198872412.003.0007

The stakes are further raised when it is realized that—in academia as well as in politics—all these dimensions are open to deliberate or inadvertent bias by various kinds of privilege and power. Here as elsewhere in any kind of democracy, the most crucial countervailing quality is openness. This is why the insights of the Taoist philosopher Lao Tzu are so relevant in modelling, including that: 'knowing ignorance is the best part of knowledge' (Tzu 1955). To paraphrase George Box (1976), such missing knowledge means 'some models may be useful, but all are wrong' (Holmdahl and Buckee 2020). For it is only through honesty, humility, and reflexivity about the inescapable presence of inconvenient unknowns that models may be truly useful. Otherwise, if employed carelessly (for example with disingenuity, hubris, or under-qualification), models are more likely to lead to deception than to enlightenment.

So, why do models continue to be used so ubiquitously and so assertively in policy-making, without due care or qualification? Why are rhetorics of 'sound science' (Wagner and Steinzor 2006) and 'evidence based policy' (Prewitt, Schwandt, and Straf 2012) becoming ever more strident—giving impressions (in areas like climate disruption, nuclear safety, biodiversity loss, and toxic health risks) that modelling can form a sufficient—even definitive—basis for policy decisions? Why (for instance in referring to International Panel on Climate Change (IPCC) models) do even critical social movements, formerly well aware of the flaws, blinkers, and imprints of power in scientific modelling, now mobilize around slogans like 'do what the science says' (Stirling 2020)?

This growing gap between the scientific and political realities around modelling is odd. After all, it is not as if history has not repeatedly demonstrated the persistent presence of ignorance. The ozone hole arose from chemical reactions not initially included in regulatory models (Harremoës et al. 2001). Before they were recognized to be causing the industry-devastating BSE (bovine spongiform encephalopathy) epidemic, food-safety modelling also effectively ignored a whole class of potential 'prion' pathogens (Millstone and van Zwanenberg 2005). Likewise, mechanisms of harm associated with endocrine-disrupting chemicals were also for decades entirely excluded from standard-setting models (EEA 2013). With Three Mile Island, Chernobyl, and Fukushima, a grim succession of nuclear reactor accidents has occurred, of kinds that canonical nuclear safety models initially confidently insisted were of 'negligible' likelihood (Wheatley, Sovacool, and Sornette 2016).

These kinds of ever-present possibilities that unknowns will yield surprises are actually a key motivation for scientists (Gross 2010). Likewise, when aligned with prevailing commercial interests, positive 'serendipity' is keenly pursued in innovation (Yaqub 2018). So why does the perennial condition of ignorance remain so crucial yet so neglected, when scientific modelling meets political of decision-making? This chapter will illuminate some hidden forces that pressure this precarious blind spot in policy modelling—and explore some far-reaching implications.

Unknowns beyond uncertainty

At first sight, the above picture may seem a little overstated. Is it not routine, after all, that policy models do, in fact, acknowledge their uncertainties? Isn't 'evidence-based' modelling for policy usually clear about the probabilities with which to expect contrasting outcomes—and the need for statistical sophistication in how they are interpreted? Even if they are sidelined in the resulting policy summaries, aren't these problems addressed through use of tools like error bars, significance values, and standard deviations in communicating technical findings (Petersen et al. 2012)?

Sadly, any reasonable answer has to be 'no'. First, policy models are in practice all-too-often expressed without any such probabilistic qualifications (Keynes and Lewis 1922). Political forces push for 'one handed experts' to 'keep it simple' and offer a convenient 'elevator pitch' for powerful sponsors (Stirling 2010). Even where statistical methods are used at all in presentation, this can sadly often be more obscuring than illuminating of underlying challenges of ignorance (Spiegel-halter 2019). The truth is that even some of the most apparently sophisticated ways in which policy modelling can seek to address uncertainty more often compound than resolve the above concealment of unknowns.

One deep but routine way in which conventional treatment of uncertainties in modelling exacerbates dilemmas of hidden unknowns lies in apparently innocent use of language. More than a century ago, the economist Frank Knight made explicit what has always been a crucial distinction in decision-making. For him, 'a measurable uncertainty, or "risk" proper [...] is so far different from an unmeasurable one that it is not in effect an uncertainty at all' (Knight 1921). But current modelling is saturated with a tendency to treat 'uncertainty' as if it were adequately addressed by means of apparently neatly quantified probabilities (Morgan, Henrion, and Small 1990). Knight's divide between 'risk' (where probabilities are clear) and 'uncertainty' (where they are not), is routinely obscured (see also 'Developing a taxonomy of uncertainty' in chapter 10). It is because of this seriously confusing—sometimes actively manipulative—use of terminology that it has been suggested that the term *incertitude* be used to encompass the wide range of ways in which knowledge may be incomplete, so as to avoid implications that probabilities are more comprehensive and applicable than they really are.

Yet this is not just a question of language. Terminologies around the treatment of unknowns in policy-making are notoriously complex, confused, and opaque (Spiegelhalter and Riesch 2011). This is often taken as a chance to exaggerate authority and manipulate debate (Amoore 2013). But the real problems lie in the categories themselves, not their names. If it is not to be called 'uncertainty', then what other name should be given to a situation in which there is a lack of confidence in probabilities (Smith and Stern 2011)? If (as sadly is often the case), there is effectively no word for this, then a serious act of concealment has occurred. If 'uncertainty' and 'risk' are treated as effectively the same and grave questions over the validity of probabilities relegated to obscure adjective-confined categories (like 'deep' or 'strict' uncertainty), then the language is barely less deceptive (Sutherland, Spiegelhalter, and Burgman 2013). Either way, arguably the most centrally prevalent dilemma in real-world decision-making is misleadingly marginalized, as if it were (at best) just a secondary side issue or (at worst) non-existent.

As if this were not grave enough, there is another—related but distinct—way in which unknowns are routinely hidden in modelling. This involves concealment of what are often very large-scale sensitivities and variabilities in the results produced by policy models (Shackley and Wynne 1996). Differing equally reasonable model input terms or parameter structures will typically yield seriously contrasting but equally rigorous results (Lauridsen et al. 2002). Spanning techniques like risk assessment, life-cycle analysis, environmental valuation, ecosystem services, cost–benefit appraisal, and decision theory, it is commonplace that model findings for particular decision options vary by several orders of magnitude (Stirling 1997). If fully described, the uncertainty intervals for contrasting policies often overlap so seriously that even relative orderings are effectively indeterminate (see chapter 5).

What this means is that it is not just in theoretical terms that modelling can be seen to be routinely unable to provide singular definitive grounds for particular decisions. Where there is openness, this predicament is also evident in the practical 'empirical' picture typically yielded in model outputs. But such transparency is rare. Instead, modellers, their sponsors, and communicators more usually remain publicly silent about the full envelope of valid outputs and highlight instead just a specific subset of selected 'results'—often asserted with fine degrees of precision that implicitly convey misleadingly high levels of confidence. Where these realities of modelling are ignored or suppressed, policy debates can remain not only seriously underinformed about real model uncertainties, but actively deceived (Stirling 2010).

Suppressed ambiguities as strategic unknowns

It is not just the ways probabilities are treated that make policy modelling vulnerable to the suppression of unknowns. A further key problem is that the real world addressed by any given model is typically multidimensional in many other ways. Too often, policy demand for models takes it for granted that there is a particular single dimension that is self-evidently of overwhelmingly greater interest than others. This can then be treated as if it were the only parameter that mattered, with other dimensions (that may actually be well understood and highly prioritized by many), being artificially neglected and so effectively rendered unknown.

For example, under rules mandated by the World Trade Organization (WTO) for food risk models, a key focus is consumer safety, rather than many other kinds of benefit or harm to citizens, farmers, workers, economies, animals, wider nature, or democracy (Fisher 2006). Likewise, standard setting for household products or chemicals is routinely confined to health risks, rather than broader values (van Zwanenberg and Millstone 2000). Modelling in such fields is not usually (as often supposed) about comparing what might broadly be the 'best' course of action, but instead about asking whether risks from a particular powerfully favoured decision might be seen as merely 'tolerable' (Bouder, Slavin, and Löfstedt 2007). Whether such risks may be made *more* tolerable by different decisions, whether they are justified, whether claimed associated benefits are real, or how pros and cons of different policy alternatives might be distributed, all often remain beyond the scope of modelling. This is how policy models can make invisible (and so unknown) issues that might be expected to be central (Stirling 2010).

Even when policy modelling does not exclude key dimensions of interest like this, it can still actively construct unknowns in other ways (see 'The numbers critique' in chapter 6). To recognize this does not require any external critique. It is much to the credit of the disciplines of axiomatic logic underlying the mathematics of modelling that one further key challenge has been repeatedly well explored in this field (Kelly 1978). Indeed, foundational work in rational choice theory (which is well known in the profession, but largely unknown outside), shows an important difficulty arising not just in the excluded dimensions discussed above, but in terms of the calculations themselves (Arrow 1963). This problem is that models reduce multidimensional realities to lower-dimensional representations. This necessarily involves aggregating values across incommensurable measures. No matter how narrow the scope, a classic problem of 'apples and oranges' is thus always present in modelling for policy—for instance across environmental, health, economic, or ethical considerations, or across different kinds of risk. In a plural society, it is a matter of Nobel Prize-winning logic that there can be no guarantee that even relative orderings of different policy choices will be determinable (MacKay 1980). If (as is normal) modelling focuses on results under some subset of possible weightings and leaves others unexplored, then it is again effectively rendering unknown the implications of alternatives to the chosen value judgements (Stirling 2003).

What this means is that the ostensibly precise numerical values often produced as outputs in policy models are also problematic for additional reasons beyond those discussed above. Not only do simple scalar numbers miss entire dimensions of concern; they also mislead about how, and with what implications, those dimensions that *are* included in a model can be differently prioritized. In these terms, even a relatively straightforward state of Knightian 'risk' can become deeply intractable. Probabilities themselves may be thought to be confidently derivable, but the categories of 'benefit' or 'harm' to which they apply may still be radically ambiguous and contestable (Rayner and Cantor 1987). As attention extends beyond risk to ambiguity, uncertainty, and ignorance, it grows increasingly clear that policy modelling is a highly selective performance (Klinke et al. 2006). Indeed, the kinds of apparently singular, robustly prescriptive findings routinely aspired to—and often claimed—in policy modelling, become recognizable as a pretence. Again, policy models are as important in the unknowns they conceal as the knowns that they highlight.

Models make unknowns as well as knowns

What all this reminds us of is that 'the unknown' is not a condition '*out there*' in the world: it is a state of knowledge (Stirling 2019b). Like all knowledge, what is unknown is deeply embedded in—and shaped by—society. This is as true in modelling as it is in wider life. Realizing this could hardly be of more profound practical importance, especially in today's turbulent world—for the many ambiguities, gaps, exclusions, and suppressions of unknowns in policy modelling are not always innocent. Just as 'knowledge is power', so too is the shaping of non-knowledge (Gross and McGoey 2015).

In one sense, this message about the political salience of unknowns in modelling may actually be thought to be widely acknowledged. Instances of deliberate covert manipulation of policy models have been well documented and intensively discussed over many years—for instance around tobacco, chemicals, pharmaceuticals, biotechnologies, automobiles, nuclear energy, and fossil fuels (Ioannidis 2007). Multiple cases emerge in which 'merchants of doubt' (Oreskes and Conway 2010) 'corrupted' (Ravetz 2016) regulatory science into overt 'dishonesty' (Lewontin 2004) in order to 'manufacture ignorance' (Proctor and Schiebinger 2008) about the adverse effects of their activities and to 'engineer consent' (Bernays 1955) for their favoured products and policies. In this way, powerful industrial interests have been able to subvert particular modelling programmes in ways that clearly breach principles of rigour and accountability—capturing regulatory institutions, compromising ostensibly independent academic expertise, and warping wider public debate.

To be fair, however, a range of widely advocated conventional remedies are also well known around such cases. Typically expressed as means to restore 'trust in science' after episodes in which such deceptions have come to light, these involve reassertion in regulatory modelling of scientific qualities that are traditionally publicly aspired to like independence, integrity, rigour, and transparency—as well as greater public oversight and separation of promotional and regulatory functions (ALLEA 2019). Yet what is often less well known is that the policy implications of hidden unknowns in modelling go far beyond these individual lurid examples and their conventional fixes. For none of the fundamental processes described in this chapter, by which modelling can conceal unknowns, necessarily require any deliberate regressive intent. As the above examples show, the nature of routine policy modelling does provide fertile ground for serious deliberate manipulation. But other kinds of equally serious warping effect can occur, even in the absence of malign aims. So, many ways in which conventional policy models quietly suppress some unknowns and create others are 'more cock up than conspiracy'. Within the bounds of rigour and integrity, policy modelling typically offers so much latitude for equally 'reasonable' divergent designs and interpretations, that it is hardly surprising that political pressures lead interests involved to highlight those results that are more expedient and leave out those that are inconvenient.

A telling example here can be found in the longstanding and high-profile practice of 'externality assessment': the calculation of the costs incurred by society as a whole due to the wider social and environmental effects of particular products or technologies (Pearce and Turner 1990). As seen in the prime example of the 'carbon tax', the idea is that regulatory measures can then add these carefully modelled 'external costs' onto prevailing market prices, to ensure that consumption decisions take reasonable account of the negative impacts and elevated prices result in less use of damaging options (Rennert et al. 2021). When these monetary 'adders' are introduced into legislation, they are typically expressed (for reasons discussed earlier) in very precise terms—without any uncertainty or ambiguity. But (again consistent with the concealment of unknowns discussed above) the background peer-reviewed literature shows that the selected values for these adders might easily—and equally rigorously—have differed by several orders of magnitude (Stirling 1997). Given this latitude, what often occurs is that those 'modelling results' that are selected for formal policy instruments are monetary values that fall within a narrow band defined by prevailing wholesale prices. This is because both policy makers and modellers know that, if numbers identified for adders are too small, they will have no impact on market decisions. But if these numbers are too high, then they will have a crippling effect on existing markets. This tendency for highlighted externality model outputs to fall within a very narrow range (irrespective of the wider scientific variability) has been termed the 'price imperative' (Stirling 1997). With no necessary regressive intent or bias towards particular options, but simply in hope of being helpful to policy, modelling is again complicit in expedient concealment.

Crucially, what this also illustrates is the neglected reality that models are typically as much about producing unknowns as about building knowledge. In fact, it is arguable that neither can take place without the other. On the face of it, policy modelling certainly produces and organizes knowledge that can inform better decisions. Much expertise, effort, and good faith is expended in seeking to ensure that these aims are addressed (van Beek et al. 2022). Yet at the same time, the people and processes involved can—inadvertently and invisibly—become trapped on politically conditioned 'gradients of expediency' which shape the understandings that they are drawn to highlight and the recommendations they feel obliged to assert (Stirling 2019a). Here, one example can be found in the field of nuclear reactor risk modelling, mentioned earlier. Historically in Japan, anyone with common sense knew very well that particular low-lying coastal districts were vulnerable to tsunamis. Medieval inscribed stones on hillsides reminded people of the heights reached by former disasters (Lewis 2015). Yet detailed risk models calculated that nuclear power plants like Fukushima could safely be built on such sites. In the process of performing these calculations, however, a wide range of relevant factors were not considered. As a result, the models gave answers that were expedient to their sponsors and yet which gravely underestimated the risks (NAIIC[1] 2012). What effectively happened was that production of ostensibly improved knowledge through modelling actually rendered effectively unknown some crucial truths that were previously well understood—that is, one should not make vulnerable commitments in these areas.

Modelling as closure for justification

All this returns to the point from George Box mentioned at the beginning. Even when they are 'wrong', policy models can indeed be 'useful' (Box 1976). The key questions in any given instance will be: what exactly is this use? For whom? To what end? And here again, what is unspoken may often be most important. So it is as necessary to be realistic and rigorous in scrutinizing the various motivations, pressures, and incentives operating around policy modelling as it is to be critical in checking model design, data, and results. Not all is always what it seems. A model may be presented as a way to 'speak truth to power' (Wildavsky 1987). But (even if unintentionally), it can at least as often be the case that this model also serves as a means by which 'power shapes truth' (Stirling 2015). Of course, real-world cases lie on a rich spectrum between such divergent kinds of 'use'. So—rather than naïve assumptions, romantic delusions, or outright denial of this unavoidable politics— what is needed is open, reasoned debate over the different political uses of models.

Across the many different perspectives, contexts, and orientations seen in the political use of modelling, one basic dynamic comes to the fore. This concerns the strong pressures that can arise, on all sides, to exaggerate the authority of a chosen body of knowledge as a means to 'close down' space for political contestation (Stirling 2008). Long studied across the social and political sciences, this syndrome includes many different forms of what is often known as *justification* (Boltanski

---

[1] The National Diet of Japan Fukushima Nuclear Accident Independent Investigation Commission.

and Thévenot 1991). In one form that has been called (Stirling 2008) 'strong justification', the aim is to support a particular direction for action that is known in
advance (Collingridge 1980). In 'weak justification', on the other hand (Stirling
2008), the effort to secure closure is more general, seeking to reinforce broader
institutions or interests and whatever actions they may later seek to undertake
(Habermas 1976). Either way, these wider insights are crucial to understanding
the many neglected features and uses of policy models discussed above.

Although rarely directly acknowledged in the presentation of policy modelling,
many of these uses are inadvertently betrayed by the preoccupations of associated policy literatures. For instance, one common motive for closing down
design, implementation, or communication of models lies in the political value
of the much-discussed quality of *public acceptance*. Is this more about truth or
domination (Wynne 1983)? Another motive lies in the hope that an exaggerated
impression of confidence will foster greater *public trust* for those interests shaping
the model (Löfstedt 2005). However, what about trust or sceptical publics from
the disciplines and interests behind the modelling (Stirling 2015)? A third driver
is an ever-present pressure in real-world politics for *blame management* (Hood
2011). By delivering a spuriously precise and over-confident rationale for decision
makers, are not modellers becoming highly partisan?

These worldly imperatives act together to create an apparently Kafka-esque
everyday reality in policy modelling. The typically artificially 'closed down' picture
of knowledge favoured by prevailing cultures around modelling can be more valuable to incumbent power and privilege as a means to justify actions (even if they
are seen to go awry) than as a source of confidence that they will not go wrong
in the first place. In other words, in the politics of modelling (as more widely),
justification can be a more precious commodity than truth.

One benefit in more clear-eyed reflection about these neglected politics of models lies in realizing that it need not always be seen as self-evidently negative, that
presentation of modelling results may accentuate some particular subsets of valid
findings rather than others. As long as principles of rigour have been respected,
reasonable views on such tendencies partly (but to an important degree) depend
on which side one is on. For researchers seeking to pressure greater climate action,
for example, it will seem legitimate to emphasize 'scientific closure' around this
interest (Lahsen 2005). Likewise, for bioscientists committed to the possible benefits of gene-edited crops, a favourable interpretation can usually be constructed
within the wide array of alternative pictures accommodated by the available modelling of potential human, agronomic, or ecological impacts (Baulcombe et al.
2009). In the same analytical spirit as modelling itself, the key reason for considering these inescapable features of the political environments of modelling is
not to judge normatively whether any given instance is 'bad' or 'good', but to
avoid a blindness to often-important consequences that would otherwise remain
unknown.

It is in this light that it is crucial, both in scientific and democratic terms, to openly recognize the importance of the latitude for debate afforded by modelling uncertainties, ambiguities, and unknowns (Saltelli et al. 2020a). This is how room for reasonably diverging emphasis in interpretations can be distinguished from the outright misrepresentations, manipulations, and concealments attempted by some of the industrial interests mentioned above, which breach principles of scientific rigour or democratic accountability. The differing interpretations that a model will typically support are broad and not bounded in any simple way. But they are nonetheless not unlimited. So, acknowledging space for a legitimate plurality of equally valid policy recommendations is very different from both dogmatic assertions of a singular 'truth' and a caricature 'relativist' position in which 'anything goes' (Stirling 2015). The plurality of legitimate understandings typically yielded in the hidden variabilities of modelling may be destabilizing, but it is nowhere near infinity. So what is just as important as the recognizing of this *plurality*, is the exploring of the *conditions* associated with each relevant interpretation (see chapter 8). It is around this need for *plural and conditional* use of policy models that arguably the most crucial concrete practical conclusions arise (Stirling 2008).

Opening up modelling for rigour and democracy

So, how in practice should we address, positively and progressively, the grave challenges summarized in this chapter? Does the modelling baby need to be thrown out with the justificatory bathwater? Or might a more caring and less controlling response lie in recognizing the synergies and balances and entanglements explored earlier—between science and democracy, closure and openness, knowns and unknowns (Scoones and Stirling 2020)? In particular, how might policy modelling escape the reductive-definitive straitjacket in which it is currently held hostage, and become more *plural* and *conditional* in ways introduced above (Stirling 2008)? *Plural* in the sense of avoiding singular truth claims and instead acknowledging multiple valid interpretations. *Conditional* in rigorously and accountably stating the determining conditions for each (see 'The model of everything' in chapter 3).

Here, there is (perhaps surprisingly) no shortage of concrete, pragmatic, and straightforward means that may serve these ends in diverse ways across contrasting contexts. Again, there is a further methodological sense in which tools that are well known to specialists are being rendered effectively unknown in wider debate. Either way, one key initial step (whose importance ought not be underestimated), is to harness the catalytic effect of both experts and policy makers openly acknowledging that models are not just a matter of 'speaking truth to power', but also unavoidably about 'power shaping knowledge' (Stirling, Ely, and Marshall

2018). This then becomes not a pathology to be decried, accused, and denied, but a reality to be recognized and cared for. To admit this might help defuse polarizing 'sound science' dogma in expressing modelling results. This may in turn help nurture more nuanced forms of questioning in wider politics—further supporting demands for more plural and conditional modelling.

Related to this is the need to actively challenge, at every mention, the currently overbearing forms of rhetoric around models as 'science-based policy' or 'evidence-based decisions' (Saltelli et al. 2022). Indeed, where these terms imply a unique basis for some singular decision, they should be taken as diagnosing exactly the opposite dynamics at work: concealing the shaping roles of power-in-knowledge. Here it is a responsibility born of rigour to point out that such aims and claims are not just undermining of democracy but also of science (Saltelli 2018). Instead, the open-endedness of modelling might be better expressed by referring to 'science-informed policy' (Amendola 2001) and 'evidence bound policy' (Stirling 2005).

Beyond this, modelling itself has over the years developed many more specific, well-tested operational procedures that can help explore the essential pluralities and conditionalities discussed in this chapter (Walker et al. 2003). Here, a particular value may be found in hybrid quantitative/qualitative and 'transdisciplinary' methods (Stirling 2006). These point out in concrete terms how an ostensibly 'pure' quantitative idiom in conventional policy models does not of itself make the models objective, still less comprehensive. By explicitly highlighting the qualitative dimensions, hybrid methods of many kinds can give cues for critical questions and spurs to developing alternative interpretations (Alasuutaru, Bickman, and Brannen 2008). Recognizing the ever-present importance of alternative categories and relations, they show that all 'quantitative problems' are always also fundamentally qualitative at root.

Examples of such 'Cinderella methods' include (see chapter 8): sensitivity auditing (Saltelli et al. 2020b), interval analysis (Mustajoki, Hämäläinen, and Lindstedt 2006), scenario processes (Derbyshire 2017), uncertainty heuristics (Halpern 2003), interactive modelling (De Marchi 2000), exploratory modelling (Bankes 1993), backcasting workshops (Robinson 2003), q method (Stephenson 1953), multi-criteria mapping (Stirling and Coburn 2018), post-normal science (Funtowicz and Ravetz 1990), flexibility assessment (Collingridge 1983), diversity analysis (Stirling 2007), precautionary appraisal (Renn et al. 2009), participatory deliberation (IIED 2003), transformation labs (Ely 2021), a pathways approach (Leach, Scoones, and Stirling 2010), reflexive design (Grin, Felix, and Bos 2004), participatory design (Schuler and Namioka 1993), open appraisal (Ely, van Zwanenberg, and Stirling 2013), civic research (Bäckstrand 2003), and statactivism (Bruno, Didier, and Vitale 2014). Not pretending at single definitive results, these approaches systematically explore conditions yielding different reasonable answers.

Crucially, this is not just a familiar injunction to 'public engagement'. Indeed, where undertaken instrumentally (for example emphasizing single 'verdicts' or 'consensus'), 'inclusive deliberation' can raise problems of spurious closure just like those in modelling. So, even where methods like the above are not themselves directly participatory, the plural and conditional nature of their outputs may help highlight necessity—and so create demand—for further interrogation of modelling design and inputs in light of public understandings, interests, and values (Stirling 2008). What matters, then, is not the claims made for any individual method as a 'panacea', but a mutually challenging ecosystem of approaches, in which quantification is about systematic exploration rather than aggregation and reduction—and always subject to qualitative questioning (Chilvers and Kearnes 2016).

Conclusion: Modelling and mutually reinforcing authoritarianisms

The argument of this chapter is readily summarized. Alongside many useful insights, there are also many ways in which the routine practice of policy modelling serves to conceal some important unknowns and create others. This should not be taken as criticism: it is simply how things are. Such closure is not self-evidently bad; it is how power and privilege of all kinds work in everyday policy-making (Rayner 2012). Implicating every side in the controversy, pressures to unrealistically exaggerate the authority of decision-making are an inherent feature of the politics of knowledge. Whether any given case is seen as justified or not will partly depend on the views in question and the values and interests at stake. But what is more clearly negative from any perspective—and as much in any normative terms as in analytic ones—is that these inescapable realities of policy modelling themselves remain underacknowledged by policy makers, underexplored by modellers, and underdiscussed in wider debate. For these crucial truths themselves to be thereby rendered unknown betrays both science and democracy (Stirling 2008). And this is arguably a historical moment, in which it is especially important to defend both (Heazle and Kane 2016).

So here at the end, attention can usefully focus on relations between policy modelling and wider democracy (Felt et al. 2008). Too often, debates on these issues are pervaded by expedient caricatures which suggest that scientific qualities and democratic values are somehow in tension. Assertive overconfidence is often confused with expertise. Quantification is mistaken for rigour. Dogmatic authority is taken for rationality. Exaggerated precision is conflated with truth. Science is idealized as something free of politics or values. Public reason is misrepresented as irrational emotiveness. When directed at the institutions and interests behind policy modelling, well-founded ('scientific') qualities of societal scepticism are treated as a general pathology. Democracy itself (in all its vibrantly contestable forms and implications) can even—very dangerously—be branded a negative impediment. For instance, in model-dominated fields like climate action, it is disturbingly asked whether 'it may be necessary to put democracy on hold' (Hickman 2010).

These syndromes too are explicable in light of the analysis offered here, as evidence of the active shaping of what is both known and unknown in such ways as to be convenient to incumbent privilege and power (Scoones and Stirling 2020). After all, whatever detailed political forms are taken by democratic struggle of different kinds, it is to dominant interests that it is often the greatest threat: as 'access by the least powerful to the capacities for challenging power' (Stirling 2019a). And it is in this light that a crucial synergy and commonality can be recognized between the historical culture of science that underlies modelling and the wider emancipatory imperatives motivating democracy in this broadest of senses (Callon, Lascoumes, and Barthe 2001). For, in terms of the politics of knowledge addressed in this chapter, science itself can be seen as a capacity for challenging exactly the kinds of power-in-knowledge discussed in this chapter (Jasanoff 2003). As recalled in the seventeenth-century motto of the Royal Society, *nullius in verba* (not on authority), a distinctive aspiration in science as a social movement, lies (for all its failings) in its aims to uphold open scepticism over closed dogma (Royal Society 2008). Yet, especially in and around the practices of policy modelling, the political pressures discussed in this chapter are increasingly leading science itself to be paraded in particular fields, less as a process of questioning, and more as a body of doctrine. After four hundred years, history may be turning full circle—science is moving from a process for challenging power-in-knowledge to becoming a privileged and apparently rigid body of dogma in its own right (Stirling 2015).

And here there emerges a final, possibly momentous implication for contemporary global politics. It has been remarkable over recent years how formerly fragmented partisan (often mutually antagonistic) nationalisms have become synchronized in a worldwide turn towards authoritarian populism—for instance in the US, China, India, Russia, Brazil, Turkey, and the UK (Norris and Inglehart 2019). This is the context for the burgeoning 'post-truth', 'plague of experts' demagoguery that is so lamented in precisely the discourses that most appreciate 'science-based' approaches to policy modelling (Bronk and Jacoby 2020). Associated formerly globalizing liberal progressive cultures and institutions, on the other hand—those bent on advocating the international rationality of 'sound scientific' governance—have ironically (despite their cosmopolitanism) failed to consolidate similarly expansive patterns (Scoones and Stirling 2020). What might explain this apparent paradox? How might it be challenged?

Strangely, the tendencies in modelling discussed here, towards concealing some unknowns and engineering others, may be central to these questions. For one way to understand the remarkably concurrent global expansion of 'charismatic authoritarianism' is as a reaction to worldwide trends over past decades towards technocratic 'expert authoritarianism' (Crate and Nuttall 2016). Behind both kinds of spurious assertiveness can be seen a similar dynamics of justification—in the overbearing truth claims, ignorance denial, and manufactured unknowns that can be (as we have seen) so expedient to entrenched-but-concealed power. Albeit inadvertently, then, the WTO and IPCC (as organizations featuring in the above account) are just two examples of major institutions that have become routinely involved in neglecting how their own favoured models are as much about propagating expedient unknowns as about producing (what is to them) useful knowledge.

Perhaps one way to understand the conundrum around the rise of authoritarian populism, then, is as a reaction to this shift of register in the global consolidation of power. Had history unfolded differently, it might have been the political left that identified this oppressive knowledge politics, in which scientific institutions allow themselves to become implicated. Perhaps due to romantic rationalism, this did not happen. So it turns out instead to be the political right that most points a long-overdue finger. Either way—for all the scientistic dazzle of his paraded array—the aspiring global emperor looks embarrassingly unclothed (Saltelli and Reinert 2023).

It is in this sense that 'minding the unknowns' in policy modelling might lie right at the heart of some of the most momentous contemporary global political challenges. Rather than doubling down on current trends towards globalizing scientistic technocracy, less might be more. Approaches to policy modelling that reduce their own authoritarian tendencies may avoid provoking an alternative authoritarian backlash. Through greater restraint and humility in spurious claims to the authority of science, the countervailing post-truth reaction might itself be less provoked (Stirling 2015). By addressing the specific problems of policy modelling discussed here, it might be possible to defend both science and democracy (Jasanoff 2005). By denying 'keep it simple' justifications yearned for by incumbent interests to defend against democratic challenge, greater space might be gleaned for exploring alternative pathways of change (Scoones, Newell, and Leach 2015). 'Plural and conditional' methods may serve as 'Trojan horses' (Stirling 2016) and 'political "judo"' (Bruno, Didier, and Prévieux 2014) to help destabilize controlling (singular, reduced, aggregated) modelling styles. As in much politics, where 'the medium is the message', such methods may help prefigure and nurture more open and reflexive policy discourses, institutions, and cultures (Scoones and Stirling 2020). By thereby 'keeping it complex', globalizing power may be denied a potent means to assert its hegemony (Stirling 2010). Precious political space might be created for less privileged voices to also interrogate, influence, and interpret policy modelling—potentially helping to nurture and enliven and renew democratic struggle towards opening up a more diverse array of possible progressive futures.

References

Alasuutaru, P., L. Bickman, J. and Brannen, eds. 2008. *The Sage Handbook of Social Research Methods*, Sage, London.

All European Academies, ALLEA. 2019. *Trust Within Science: Dynamics and Norms of Knowledge Production*, ALLEA, Berlin.

Amendola, A. 2001. 'Recent Paradigms for Risk Informed Decision Making'. *Safety Science, 40*, 17–30.

Amoore, L. 2013. *The Politics of Possibility: Risk and Security beyond Probability*, Duke University Press, Durham, NC.

Arrow, K.J. 1963. *Social Choice and Individual Values*, Yale University Press, New Haven.

Bäckstrand, K. 2003. 'Civic Science for Sustainability: Reframing the Role of Experts, Policy-Makers and Citizens in Environmental Governance'. *Global Environmental Politics, 3(4)*, 24–41.

Bankes, S. 1993. 'Exploratory Modeling for Policy Analysis'. *Operations Research, 41(3)*, 432–619.

Baulcombe, D. et al. 2009. *Reaping the Benefits: Science and the Sustainable Intensification of Global Agriculture*, Royal Society, London.

van Beek, L., J. Oomen, M. Hajer, P. Pelzer, and D.D. van Vuuren. 2022. 'Navigating the Political: An Analysis of Political Calibration of Integrated Assessment Modelling in Light of the 1.5° C Goal'. *Environmental Science & Policy, 133*, 193–202.

Bernays, E. 1955. *The Engineering of Consent*, University of Oklahoma Press, Norman, OK.

Boltanski, L. and L. Thévenot. 1991. *On Justification: Economies of Worth*, Princeton University Press, Princeton, NJ.

Bouder, F., D. Slavin, and R.E. Löfstedt. 2007. *The Tolerability of Risk: A New Framework for Risk Management*, Earthscan, London.

Box, G.E.P. 1976. 'Science and Statistics'. *Journal of the American Statistical Association, 71(356)*, 791–9.

Bronk, R. and W. Jacoby. 2020. *The Epistemics of Populism and the Politics of Uncertainty*, LSE 'Europe in Question' Discussion Paper Series, London School of Economics, London.

Brooks, H. 1986. 'The Typology of Surprises in Technology, Institutions and Development'. In *Sustainable Development of the Biosphere*, edited by W.C. Clark and R.E. Munn, International Institute for Applied Systems Analysis, Cambridge University Press, Cambridge, pp. 49–72.

Bruno, I., E.E. Didier, and T.T. Vitale. 2014. 'Statactivism: Forms of Action between Disclosure and Affirmation'. *Partecipazione e conflitto, 7(2)*, 198–220, doi: 10.1285/i20356609v7i2p198.

Bruno, I., E. Didier, and J. Prévieux. 2014. *Statactivisme. Comment lutter avec des nombres*, Zones, La Découverte, Paris.

Callon, M., P. Lascoumes, Y. and Barthe. 2001. *Acting in an Uncertain World: An Essay on Technical Democracy*, The MIT Press, Cambridge, MA.

Chilvers, J. and K. Kearnes. 2016. *Remaking Participation: Science, Environment and Emergent Publics*, Routledge, London, doi: 10.4324/9780203797693.

Collingridge, D. 1980. *The Social Control of Technology*, Frances Pinter, London.

Collingridge, D. 1983. 'Hedging and Flexing: Two Ways of Choosing under Ignorance'. *Technology Forecasting and Social Change, 23(2)*, 161–72.

Crate, S.A. and M. Nuttall, eds. 2016. *Anthropology and Climate Change: From Actions to Transformations*, Routledge, London.

Derbyshire, J. 2017. 'Potential Surprise Theory as a Theoretical Foundation for Scenario Planning'. *Technological Forecasting & Social Change*, *124*, 77–87, doi: 10.1016/j.techfore.2016.05.008.

EEA, European Environment Agency. 2013. *Late Lessons from Early Warnings: Science, Precaution, Innovation—Summary*, European Environment Agency, Copenhagen.

Ely, A., ed. 2021. *Transformative Pathways to Sustainability: Learning across Disciplines, Cultures and Contexts*, Routledge, London.

Ely, A., P. Van Zwanenberg, and A. Stirling. 2013. 'Broadening Out and Opening Up Technology Assessment: Approaches to Enhance International Development, Co-Ordination and Democratisation'. *Research Policy*, *43*(3), 505–18, doi: 10.1016/j.respol.2013.09.004.

Felt, U., B. Wynne, M. Callon, and M.E. Gonçalves. 2008. *Taking European Knowledge Society Seriously: Report of the Expert Group on Science and Governance to the Science, Economy and Society Directorate, Directorate-General for Research, European Commission*, edited by U. Felt and B. Wynne, European Commission, Brussels.

Fisher, E. 2006. 'Beyond the Science/Democracy Dichotomy: The World Trade Organization Sanitary and Phytosanitary Agreement and Administrative Constitutionalism'. In *Constitutionalism, Multi-Level Trade Governance, And Social Regulation*, edited by A. Joerges and E.-U. Petersmann, eds, Hart Publishing, London, pp. 327–50.

Funtowicz, S. and J.R. Ravetz. 1990. *Uncertainty and Quality in Science for Policy*, Kluwer Academic Publishers, Dordrecht.

Grin, J., F. Felix, and B. Bos. 2004. 'Practices for Reflexive Design: Lessons from a Dutch Programme on Sustainable Agriculture'. *International Journal of Foresight and Innovation Policy*, *1*(1–2), 126–49.

Gross, M. 2010. *Ignorance and Surprise: Science, Society and Ecological Design*, The MIT Press, Cambridge, MA.

Gross, M. and L. McGoey, eds. 2015. *Routledge International Handbook of Ignorance Studies*, Routledge, London.

Habermas, J. 1976. *Legitimation Crisis*, Polity Press, Cambridge.

Halpern, J.Y. 2003. *Reasoning about Uncertainty*, MIT Press, Cambridge, MA.

Harremoës, P., D. Gee, M. MacGarvin, A. Stirling, J. Keys, B. Wynne, and S.G. Vaz, eds. 2001. *Late Lessons from Early Warnings: The Precautionary Principle, 1896–2000*, European Environment Agency, Copenhagen.

Heazle, M. and J. Kane, eds. 2016. *Policy Legitimacy, Science and Political Authority: Knowledge and Action in Liberal Democracies*, Routledge, London.

Hickman, L. 2010. 'James Lovelock: Humans Are Too Stupid to Prevent Climate Change'. *The Guardian*, 29 March 2010, pp. 2–5.

Holmdahl, I. and C. Buckee. 2020. 'Wrong but Useful—What Covid-19 Epidemiologic Models Can and Cannot Tell Us'. *New England Journal of Medicine*, *383*, 303–5.

Hood, C. 2011. *The Blame Game: Spin, Bureaucracy, and Self-Preservation in Government*, Princeton University Press, Princeton, NJ.

IIED, International Institute for Environment and Development. 2003. 'Participatory Processes for Policy Change', International Institute for Environment and Development, London, p. 96, p.: ill., maps 30 cm.

Ioannidis, J.P.A. 2007. 'Why Most Published Research Findings Are False: Author's Reply to Goodman and Greenland'. *PLoS Medicine*, edited by A.M. Huberman and M.B. Miles Public Library of Science (WISICT '04), *4*(6), 2.

Jasanoff, S. 2003. 'Technologies of Humility: Citizen Participation in Governing Science'. *Minerva*, *41*, 223–44.

Jasanoff, S. 2005. *Designs on Nature: Science and Democracy in Europe and the United States*, Princeton University Press, Princeton, NJ, doi: 10.1017/CBO9781107415324.004.

Kelly, J.S. 1978. *Arrow Impossibility Theorems*, Academic Press, New York.

Keynes, J.M. and C.I. Lewis. 1922. 'A Treatise on Probability'. *The Philosophical Review*, *31*(2), 180, doi: 10.2307/2178916.

Klinke, A., M. Dreyer, O. Renn, A. Stirling, and P. van Zwanenberg. 2006. 'Precautionary Risk Regulation in European Governance'. *Journal of Risk Research*, *9*(4), 373–92, doi: 10.1080/13669870600715800.

Knight, F.H. 1921. *Risk, Uncertainty and Profit*, Houghton Mifflin, Boston, MA.

Lahsen, M. 2005. 'Seductive Simulations? Uncertainty Distribution Around Climate Models'. *Social Studies of Science*, *35*(6), 895–922, doi: 10.1177/0306312705053049.

Lauridsen, K., I. Kozine, F. Markert, A. Amendola, M. Christou, and M. Fiori. 2002. *Assessment of Uncertainties in Risk Analysis of Chemical Establishments: The ASSURANCE Project Final Summary Report*, Risø National Laboratory, Roskilde, Denmark.

Leach, M., I. Scoones, and A. Stirling. 2010. *Dynamic Sustainabilities: Technology, Environment, Social Justice, Dynamic Sustainabilities: Technology, Environment, Social Justice*, Earthscan, London, doi: 10.4324/9781849775069.

Lewis, D. 2015. 'These Century-Old Stone "Tsunami Stones" Dot Japan's Coastline', *Smithsonian Magazine*, 31 August 2015, https://www.smithsonianmag.com/smart-news/century-old-warnings-against-tsunamis-dot-japans-coastline-180956448/ (last accessed 25 March 2023).

Lewontin, R.C. 2004. 'Dishonesty in Science'. *The New York Review of Books*, 18 November 2004. https://www.nybooks.com/articles/2004/11/18/dishonesty-in-science/ (last accessed 25 March 2023).

Loasby, B.J. 1976. *Choice, Complexity and Ignorance: An Inquiry into Economic Theory and the Practice of Decision Making*, Cambridge University Press, Cambridge.

Löfstedt, R.E. 2005. *Risk Management in Post-Trust Societies*, Palgrave Macmillan, Basingstoke, doi: 10.1057/9780230503946.

MacKay, A.F. 1980. *Arrow's Theorem: The Paradox of Social Choice—A Case Study in the Philosophy of Economics*, Yale University Press, New Haven.

De Marchi, B. 2000. 'Learning from Citizens: A Venetian Experience'. *Journal of Hazardous Materials*, *78*(1–3), 247–59.

Morgan, M.G., M. Henrion, and M. Small. 1990. *Uncertainty: A Guide to Dealing with Uncertainty in Quantitative Risk and Policy Analysis*, Cambridge University Press, Cambridge.

Mustajoki, J., R.P. Hämäläinen, and M.R.K. Lindstedt. 2006. 'Using Intervals for Global Sensitivity and Worst-Case Analyses in Multiattribute Value Trees'. *European Journal of Operational Research*, *174*(1), 278–92.

NAIIC. 2012. *The Official Report of The Fukushima Nuclear Accident Independent Investigation Commission*, The National Diet of Japan Fukushima Nuclear Accident Independent Investigation Commission, Tokyo.

Norris, P. and R. Inglehart. 2019. *Cultural Backlash: Trump, Brexit, and Authoritarian Populism*, Cambridge University Press, Cambridge.

Oreskes, N. and E.M. Conway. 2010. *Merchants of Doubt: How a Handful of Scientists Obscured the Truth on Issues from Tobacco Smoke to Global Warming*, Bloomsbury, London.

Pearce, D. and K. Turner. 1990. *Economics of Natural Resources and the Environment*, Harvester Wheatsheaf, Hemel Hempstead.

Petersen, A.C., P. Janssen, J.P. van der Sluijs, and J.S. Risbey. 2012. *Guidance for Uncertainty Assessment and Communication*, 2nd edition, PBL Netherlands Environmental Assessment Agency, The Hague.

Prewitt, K., T.A. Schwandt, and M.L. Straf, eds. 2012. *Using Science as Evidence in Public Policy*, National Research Council, Washington, DC.

Proctor, R. and L. Schiebinger, eds. 2008. *Agnotology: The Making and Unmaking of Ignorance*, Stanford University Press, Stanford, CA.

Ravetz, J. 2016. 'How Should We Treat Science's Growing Pains?' *The Guardian*, 8 June 2016. https://www.theguardian.com/science/political-science/2016/jun/08/how-should-we-treat-sciences-growing-pains (last accessed 25 March 2023).

Rayner, S. 2012. 'Uncomfortable Knowledge: The Social Construction of Ignorance in Science and Environmental Policy Discourses Uncomfortable Knowledge: The Social Construction of Ignorance in Science and Environmental Policy Discourses', 5147. doi: 10.1080/03085147.2011.637335.

Rayner, S. and R. Cantor. 1987. 'How Fair Is Safe Enough? The Cultural Approach to Societal Technology Choice'. *Risk Analysis*, 7(1), 3–9.

Renn, O., P.-J. Schweizer, U. Muller-Herold, and A. Stirling. 2009. *Precautionary Risk Appraisal and Management*, Salzwasser-Verlag im Europäischen Hochschulverlag, Berlin.

Rennert, K., B.C. Prest, W. Pizer, R.G. Newell, D. Anthoff, C. Kingdon, L. Rennels, R. Cooke, A.E. Raftery, H. Ševčíková, and F. Errickson. 2021. *The Social Cost of Carbon: Advances in Long-Term Probabilistic Projections of Population, GDP, Emissions, and Discount Rates*, Resources for the Future, Washington, DC.

Robinson, J. 2003. 'Future Subjunctive: Backcasting as Social Learning'. *Futures*, 35(8), 839–45, doi: 10.1016/S0016-3287(03)00039-9.

Royal Society. 2008. *History of the Royal Society: Nullius in verba*, Royal Society, London.

Saltelli, A., D.J. Dankel, M. Di Fiore, N. Holland, and M. Pigeon. 2022. 'Science, the Endless Frontier of Regulatory Capture'. *Futures*, 135, 102860.

Saltelli, A. 2018. 'Why Science's Crisis Should not Become a Political Battling Ground'. *Futures*, 104, 85–90.

Saltelli, A., G. Bammer, I. Bruno, E. Charters, M. Di Fiore, E. Didier, W.N. Espeland, J. Kay, S. Lo Piano, D. Mayo, R. Jr. Pielke, T. Portaluri, T.M. Porter, A. Puy, I. Rafols, J.R. Ravetz, E. Reinert, D. Sarewitz, P.B. Stark, A. Stirling, J.P. van der Sluijs, and P. Vineis. 2020a. 'Five Ways to Ensure That Models Serve Society: A Manifesto'. *Nature*, 582, 482–4.

Saltelli, A., L. Benini, S. Funtowicz, M. Giampietro, M. Kaiser, E. Reinert, and J.P. van der Sluijs. 2020b. 'The Technique Is Never Neutral: How Methodological Choices Condition the Generation of Narratives for Sustainability'. *Environmental Science & Policy*, 106, 87–98, doi: 10.1016/j.envsci.2020.01.008.

Saltelli, A. and E.S. Reinert. 2023. 'Physiocracy, Guillotines, and Antisemitism? Did Economics Emulate the Wrong Enlightenment?' In *A Modern Guide to Uneven*

Economic Development, edited by E. Reiner and L. Kvangraven, Edward Elgar Publishing, Cheltenham, pp. 200–17.

Schuler, D. and A. Namioka, eds. 1993. *Participatory Design: Principles and Practices*, Routledge, London.

Schwarz, M. and M. Thompson. 1990. *Divided We Stand: Redefining Politics, Technology and Social Choice*, Harvester Wheatsheaf, New York.

Scoones, I., P. Newell, and M. Leach. 2015. 'The Politics of Green Transformation'. In *The Politics of Green Transformations*, edited by I. Scoones, M. Leach, and P. Newell, Routledge, Abingdon, pp. 1–42.

Scoones, I. and A. Stirling, eds. 2020. *The Politics of Uncertainty: Challenges of Transformation*, Routledge, London.

Shackley, S. and B. Wynne. 1996. 'Representing Uncertainty in Global Climate Change Science and Policy: Boundary-Ordering Devices and Authority'. *Science, Technology & Human Values*, 21(3), 275–302, doi: 10.1177/016224399602100302.

Smith, L.A. and N. Stern. 2011. 'Uncertainty in Science and Its Role in Climate Policy'. *Philosophical Transactions. Series A, Mathematical, Physical and Engineering Sciences*, 369(1956), 4818–41, doi: 10.1098/rsta.2011.0149.

Spiegelhalter, D. 2019. 'The Art of Statistics: Learning from Data', Pelican, London.

Spiegelhalter, D.J. and H. Riesch. 2011. 'Don't Know, Can't Know: Embracing Deeper Uncertainties When Analysing Risks'. *Philosophical Transactions. Series A, Mathematical, Physical and Engineering Sciences*, 369(1956), 4730–50, doi: 10.1098/rsta.2011.0163.

Stephenson, W. 1953. *The Study of Behavior: Q Technique and Its Methodology*, University of Chicago Press, Chicago.

Stirling, A. 1997. 'Limits to the Value of External Costs'. *Energy Policy*, 25(5), 517–40.

Stirling, A. 2003. 'Risk, Uncertainty and Precaution: Some Instrumental Implications from the Social Sciences'. In *Negotiating Change: New Perspectives from the Social Sciences*, edited by F. Berkhout, M. Leach, and I. Scoones, Edward Elgar, Cheltenham, pp. 33–76.

Stirling, A. 2005. *Comments on the DEFRA Evidence and Innovation Strategy: A Response to the Consultation Document of October 2005*.

Stirling, A. 2006. *From Science and Society to Science in Society: Towards a Framework for Co-operative Research*, Report of a European Commission Workshop, Governance and Scientific Advice Unit of DG RTD, Directorate C2, Directorate General Research and Technology Development, Brussels.

Stirling, A. 2007. 'A General Framework for Analysing Diversity in Science, Technology and Society'. *Journal of the Royal Society Interface*, 4(15), 707–19, doi: 10.1098/rsif.2007.0213.

Stirling, A. 2008. '"Opening Up" and "Closing Down": Power, Participation, and Pluralism in the Social Appraisal of Technology'. *Science, Technology and Human Values*, 23(2), 262–94.

Stirling, A. 2010. 'Keep It Complex'. *Nature*, 468, 1029–31.

Stirling, A. 2015. 'Power, Truth and Progress: Towards Knowledge Democracies in Europe'. In *Future Directions for Scientific Advice in Europe*, edited by J. Wilsdon and R. Doubleday, Cambridge University Press, Cambridge, pp. 133–51.

Stirling, A. 2016. 'Knowing Doing Governing: Realising Heterodyne Democracies'. In *Knowing Governance: The Epistemic Construction of Political Order*, edited by J.-P. Voß and R. Freeman, Palgrave Macmillan, Basingstoke, pp. 259–89.

Stirling, A. 2019a. 'How Deep Is Incumbency? A "Configuring Fields" Approach to Redistributing and Reorienting Power in Socio-Material Change'. *Energy Research & Social*, 58, 101239, doi: 10.1016/j.erss.2019.101239.

Stirling, A. 2019b. 'Politics in the Language of Uncertainty'. *STEPS Centre Blog*, 11 February 2019. https://steps-centre.org/blog/politics-in-the-language-of-uncertainty/ (last accessed 25 March 2023).

Stirling, A. 2020. 'Does the Delusion of "Climate Control" Do More Harm Than Good to Climate Disruption?' *STEPS Centre Blog*, 10 July 2020, pp. 1–14. https://steps-centre.org/blog/does-the-delusion-of-climate-control-do-more-harm-than-good-to-climate-disruption/ (last accessed 25 March 2023).

Stirling, A. and J. Coburn. 2018. 'From CBA to Precautionary Appraisal: Practical Responses to Intractable Problems'. *The Hastings Center Report*, 48, S78–S87, doi: 10.1002/hast.823.

Stirling, A., A. Ely, and F. Marshall. 2018. 'How Do We "Co-Produce" Transformative Knowledge?' *STEPS Centre Blog*, 7 February 2018. https://steps-centre.org/blog/how-do-we-co-produce-transformative-knowledge (last accessed 25 March 2023).

Sutherland, W.J., D. Spiegelhalter, and M.A. Burgman. 2013. 'Twenty Tips for Interpreting Scientific Claims'. *Nature*, 503, 5–7.

Thompson, M. and M. Warburton. 1985. 'Decision Making Under Contradictory Certainties: How to Save the Himalayas When You Can't Find What's Wrong with Them'. *Journal of Applied Systems Analysis*, 12, 3–34.

Tzu, L. 1955. *Tao Te Ching*. Signet, New York.

Wagner, W. and R. Steinzor, eds. 2006. *Rescuing Science from Politics: Regulation and the Distortion of Scientific Research*, Cambridge University Press, Cambridge.

Walker, W.E., P. Harremoës, J. Rotmans, J.P. van der Sluijs, M.B.A. van Asselt, P. Janssen, and M.P. Krayer von Krauss. 2003. 'Defining Uncertainty: A Conceptual Basis for Uncertainty Management in Model-Based Decision Support'. *Integrated Assessment*, 4, 1–20.

Wheatley, S., B. Sovacool, and D. Sornette. 2016. 'Of Disasters and Dragon Kings: A Statistical Analysis of Nuclear Power Incidents and Accidents'. *Risk Analysis*, 37(1), 99–115, doi: 10.1111/risa.12587.

Wildavsky, A. 1987. *Speaking Truth to Power: The Art and Craft of Policy Analysis*, Transaction Publishers, London.

Wynne, B. 1983. 'Redefining the Issues of Risk and Public Acceptance: The Social Viability of Technology'. *Futures*, 15(1), 13–32.

Wynne, B. 1992. 'Uncertainty and Environmental Learning: Reconceiving Science and Policy in the Preventive Paradigm'. *Global Environmental Change*, 2(2), 111–27.

Yaqub, O. 2018. 'Serendipity: Towards a Taxonomy and a Theory'. *Research Policy*, 47, 169–79, doi: 10.1016/j.respol.2017.10.007.

van Zwanenberg, P. and Millstone, E. 2000. 'Beyond Skeptical Relativism: Evaluating the Social Constructions of Expert Risk Assessments'. *Science, Technology & Human Values*, 25(3), 259–82, doi: 10.1177/016224390002500301.

PART III
THE RULES IN PRACTICE

8

Sensitivity auditing

A practical checklist for auditing decision-relevant models

Samuele Lo Piano, Razi Sheikholeslami, Arnald Puy, and Andrea Saltelli

Introduction

The *Manifesto* that offers the occasion for the present volume (Saltelli et al. 2020a) is based on five rules to improve the way models are used for policy-making and advocate for a better reciprocal domestication between models and society overall. The rules are themselves the outcome of the sedimentation of earlier checklists for model quality, foremost that provided by sensitivity auditing, with contributions from other strands of scholarship.

It would be natural at this point to ask 'What is sensitivity auditing?' The short answer would be that it is an extension of sensitivity analysis (see chapter 5). In turn, sensitivity analysis is the logical complement of an uncertainty analysis, also known as uncertainty quantification. Some definitions will be helpful here.

Uncertainty analysis: By assigning a range of variation and a distribution to uncertain model inputs and assumptions, one can generate an empirical distribution function for the output(s) of interest by running the model on samples from these distributions. The expressions 'error propagation analysis' and 'uncertainty cascade' are also used (Christie et al. 2011: 86). See also chapter 2.

Sensitivity analysis is a methodology used to ascertain which of the uncertain inputs is more influential in generating the uncertainty in the output.

These two analyses 'talk' to one another. If the output has little or no uncertainty, there is no point in dissecting its uncertainty to discover 'the culprits'. Even the extreme opposite scenario, one where uncertainty is large enough to impair the meaningful application of the model (see chapter 4), does not offer a chance for any meaningful inference. In all cases in between, discovering the responsible inputs may help alleviate the issue.

These analyses imply many choices on top of the usual assumptions linked to the construction of the model as such: one has to decide what inputs are to be taken to be uncertain, how to choose ranges and distributions (often a very expensive and time-consuming step), and what method to employ to select 'sensitive' factors.

Samuele Lo Piano et al., *Sensitivity auditing*. In: *The politics of modelling.* Edited by: Andrea Saltelli and Monica Di Fiore, Oxford University Press. © Oxford University Press (2023). DOI: 10.1093/oso/9780198872412.003.0008

It is not difficult to imagine a situation where a model is used in decision- or policy-making, and where different actors have different visions of what should be modelled and how. In this case, the entire modelling process and its conclusions, including its technical uncertainty and sensitivity analyses, could be deconstructed by contesting the choices just mentioned.

The examples provided in this chapter take this approach by exploring and bringing into the open subjective or normative elements of a modelling process. In other words, a 'technical' sensitivity analysis cannot be the end of the story when the model undergoes a regulatory audit or becomes the subject of a public debate—as we have seen for the case of models related to COVID-19 (Saltelli et al. 2020a).

This approach goes by the name of *sensitivity auditing* (Saltelli and Funtowicz 2014), which is recommended *inter alia* by the European Commission and the SAPEA (2018). Sensitivity auditing consists of a seven-rule checklist (Box 8.1):

Box 8.1 The seven-rule checklist of sensitivity auditing

1. Check against a rhetorical use of mathematics.
2. Adopt an 'assumption hunting' attitude.
3. Detect Garbage In Garbage Out.
4. Find sensitive assumptions before they find you.
5. Aim for transparency.
6. Do the right sums, not just the sums right.
7. Perform thorough, state-of-the-art uncertainty and sensitivity analyses.

The first rule is *Check against a rhetorical use of mathematics*. As already discussed in chapter 4, larger models command more epistemic authority and discourage criticism. This rule invites us to appreciate the dimension of a model in relation to both its context and purpose, and the evidence that has entered the model's construction. *Adopt an 'assumption hunting' attitude* is the second rule. Models are based on assumptions, including interpretations of the underlying systems' behaviour. Some of these assumptions come with the 'tag' of the model, and are explicit—for example, when we say, 'This is an equilibrium model.' More often, assumptions are implicit. Since the construction of a model may unfold over an extended period, the same modellers may forget them.[1] The next step of sensitivity auditing is *Detect Garbage In Garbage Out (GIGO)*. This recommendation points out the circumstances where the uncertainty associated with a mathematical prediction has been overstated (magnifying uncertainty) or

---

[1] From Millgram (2015: 29): 'normally even the people who produced the model will not remember many of the assumptions incorporated into it, short of redoing their work, which means that the client cannot simply ask then what went into it'.

understated (hiding uncertainty). The latter case is perhaps the most frequent. When minimizing and/or simplifying uncertainties, a modeller aims to show that the prediction of the model is 'crisp'. For instance, nobody is interested in a cost–benefit analysis whose distribution output spans from a high loss to a high gain with equal probabilities. This strategic behaviour—'where the uncertainty of the inputs must be suppressed, lest they render its outputs totally indeterminate'—was named GIGO by Funtowicz and Ravetz (1990: 6). Reluctance to face uncertainty is a well-known issue in evidence-based policy (Scoones and Stirling 2020). The opposite gamble may be embraced to increase the uncertainty and, in doing so, defeat the assessment of, say, a regulatory agency (Saltelli 2018, Saltelli et al. 2022).

The fourth rule is *Find sensitive assumptions before they find you*. Modellers expecting a public debate around the inference produced by their models ought to better prepare in advance by running a proper sensitivity analysis. An interesting case relating to a dispute over the cost of climate change is described in Saltelli and D'Hombres (2010). Here, one of the parties in the dispute, Nicholas Stern, resorted to a sensitivity analysis after his impact assessment had been contested. However, his sensitivity analysis appeared weak when seen through the lens of sensitivity auditing (ibid.).

Aim for transparency is the fifth rule. Black-box models may simply not play well in a public debate. The open science movement strongly advocates for transparency and availability of the model source code, and making this intelligible through appropriate documentation and comments.

The sixth rule is *Do the right sums, not just the sums right*. Though this may appear the same as rule (2), it has more to do with major political or ideological worldviews. It reflects sociologists' thinking about the so-called technologies of humility (Jasanoff 2003), whereby in the production of evidence for policy, one should be careful to identify possible winners and losers, and make sure that the concerns of the latter are not missed in the debate. This rule also reflects on how quantification can determine the policy of what is being quantified (Salais 2010) in a way that is not immediately apparent due to the purported neutrality of mathematical models. Finally, the last rule is *Perform thorough, state-of-the-art uncertainty and sensitivity analyses*. Leaving undiscussed the results of a sensitivity analysis, or presenting one that is based on an inadequate design, is still fairly common practice in modelling studies across disciplinary fields (Ferretti, Saltelli, and Tarantola 2016, Saltelli et al. 2019, Lo Piano and Benini 2022); see also chapter 5.

The epistemic background of sensitivity auditing

Sensitivity auditing takes inspiration from post-normal science (PNS) (Funtowicz and Ravetz 1993), a paradigm in science for policy that applies when facts are uncertain, values in dispute, stakes high, and decisions urgent (for example in the presence of a pandemic caused by a new virus). While PNS is not easy to synthesize, its main feature is an emphasis on the quality of the assessment process, rather than the pursuit of a truth that might be elusive in conflicted issues. In a similar spirit, PNS does not appeal to the neutrality of a unitary science, and accepts that different disciplines and accredited experts may uphold different legitimate views (Saltelli et al. 2020b).

Under this paradigm, the decision-making community needs to involve more than just the experts; it should also include investigative journalists, whistleblowers, and lay citizens affected by or simply interested in the issue under analysis. Ideally, sensitivity auditing itself is also meant for use in participatory settings, where one negotiates the worth of a model, as well as the order of worth of the various sides of a problem. By the latter it means the relative position on the issue under consideration of concerns about, for example, social justice, environmental quality, or respect for existing values and traditions (Thévenot 2022). Like sensitivity analysis, sensitivity auditing functions in both construction and deconstruction, either in building a defensible, plausible model, or in demonstrating the irrelevance or contradictions of a model-based inference one is trying to contextualize, in the spirit of statactivism (Bruno, Didier, and Prévieux 2014)— for example, when one strives to bring to the surface issues or categories that an existing quantification has made invisible.

Sensitivity auditing has similarities to an older scheme of practices for the quality of numerical information known as Numeral Unit Spread Assessment Pedigree (NUSAP), which was also inspired by the PNS paradigm. NUSAP was introduced by Funtowicz and Ravetz (1990) as a way of shedding light on the quality of numbers being used in a political setting. The first two categories are self-evident: Spread reviews the error in the value of the Numeral, and Assessment and Pedigree are supposedly the results of a process of negotiation among interested parties. Assessment is a summary judgement of the quality of the numerical evidence, and might be as simple as 'conservative' or 'optimist'. Pedigree is a judgement of the quality of the process leading to the numeral, and of the numeral's policy implications. What is important in NUSAP is not so much the scores assigned to Assessment and Pedigree, but the occasion these categories offer us to engage in a reflection on the worth of the numeral (van der Sluijs et al. 2005).

The present volume has already mentioned the expanding discipline of sociology of quantification (Mennicken and Espeland 2019, Popp Berman and Hirschman 2018, Mennicken and Salais 2022). Sensitivity auditing was not developed by sociologists, yet the approach reflects many tenets of this discipline: sociologists of quantification investigate how different mechanisms of quantification are performative in shaping the reality that they should supposedly just be measuring. For this reason, one should assess a given quantification both in terms of its technical quality and its fairness. Disparate and important dimensions such as higher education (Espeland and Sauder 2016), employment (Salais 2010), and inequality (De Leonardis 2022) come to be as a result of a digitization to a large extent subtracted from democratic deliberation. While we cannot enter into this discussion in more depth here, we note that sensitivity auditing addresses this 'fairness' dimension of quantification for the case of mathematical models, and forms of quantification in general.

Finally, sensitivity auditing is one possible approach to the 'modelling of the modelling process' (Lo Piano et al. 2022, Puy, Lo Piano, and Saltelli 2020). To illustrate what this means, we need to call upon novelist Jorge Luis Borges and his short story, 'The Garden of Forking Paths' (Borges 1941). The story has been taken up in statistics (Gelman and Loken 2013) to allude to the many degrees of freedom offered to an analyst marching towards an inference. Additionally, feeding the same data to different teams can return varied statistical results (Breznau et al. 2021), and it is only natural that the effect is also seen in mathematical modelling (Refsgaard et al. 2006). Modelling of the modelling process implies, like in Borges' short story, taking all the paths instead of just one, and exploring the consequences of these different trajectories on the inference.

The next section illustrates how sensitivity auditing can be applied to models, indicators, and metrics. We will target instances of quantification used in policy-making and/or those that have received vast media coverage. Although our illustration proceeds by taking one rule at a time for each case, it should be clear that the rules can be recursive, that their order can change, and that to some extent rules overlap with one another.

The Programme for International Student Assessment (PISA)

The PISA test was designed by the Organisation for Economic Co-operation and Development (OECD) to measure the problem-solving skills of 15-year-old school pupils. The test has been undertaken every three years since 2000. Test performances on a country basis are ranked and benchmarked against a standard. In the 2018 round of the test, 79 countries participated. Criticisms of the methodological and ideological stances of the test, among other aspects, were raised in the press (see Araujo, Saltelli, and Schnepf 2017) and quoted references can be seen in chapter 9.

Several dimensions emerged after applying the sensitivity auditing checklist to the PISA test score.

> *Check against a rhetorical use of mathematics*: A reading of the test results that emphasizes a causal relation from the test scores to economic growth has been offered by prominent political bodies, including the European Commission. A document released stated that if European Union (EU) countries could significantly increase their PISA score, this would lead

to a quantifiable growth in Gross Domestic Product (GDP) for the EU (Woessmann 2016).

Adopt an 'assumption hunting' attitude: The PISA test builds on the postulate that the skills students require to succeed in current knowledge societies can be benchmarked against a one-size-fits-all international standard. Differences in curricula between countries are suppressed and excluded from the test. However, countries' wellbeing and success may emerge from these very curricular differences.

Detect pseudo-science: When communicating the test results, the survey organizers report only the standard error of the countries' test scores, neglecting important sources of volatility in country rankings. These sources of volatility may emerge, for instance, by excluding students with special educational needs or newly arrived immigrants. In past editions of the test, excluding these groups resulted in discarding more than 5% of the potential participants from some countries. However, this 5% is above the threshold imposed by the test designers as an assurance of its representativeness (Wuttke 2007). Even the tendency of less capable students to refrain from participating in the test may result in a non-representative participating cohort, thus potentially producing a bias that goes well beyond the standard error of the country's rank (Micklewright, Schnepf, and Skinner 2012).

Find sensitive assumptions before they find you: Calls for testing the sensitivity of the PISA rankings and the volatility they bring to modelling assumptions, data collection method, and use of the data items have gone unanswered (Micklewright and Schnepf 2006).

Aim for transparency: Lack of data availability represents one of the main limitations of the PISA test. This impedes an analysis of the sensitivity of the achievement scores to the modelling choices and data items used in attributing scores to countries.

Do the right sums, not just the sums right: The PISA test was conceived for the purpose of facilitating a measurement of the degree to which the teaching that students receive is useful in terms of the life challenges they may encounter in today's knowledge societies. In this sense, the test *de facto* makes the 'economic' case for education, which is framed exclusively as a means to economic growth rather than to *Bildung* and emancipation.

Perform thorough, state-of-the-art uncertainty and sensitivity analyses: Thoroughly assessing all the sources of uncertainty in the PISA rankings would require their simultaneous activation. The resulting uncertainty could produce more volatile rankings, which would require adequate communication: for example, 'uncertainties being taken into consideration, the rank of country X could vary between five and twenty' (Araujo, Saltelli, and Schnepf 2017). However, this analysis is missing from the communication on the significance of the produced countries' rank.

Nutrition and public health economic evaluations

Non-communicable diseases (NCDs) are epidemiologically affected by lifestyle habits, including diet, smoking, and physical activity level. For this reason, it is of primary importance that we evaluate how policies aimed at triggering changes in these factors may produce societal consequences in terms of disease likelihood. These policies are, in turn, informed by the available body of evidence and modelling activities. Lo Piano and Robinson (2019) identified the following criticalities:

Check against a rhetorical use of mathematics: In the field of nutrition and public health economic evaluations, an existing tendency is to resort to overly complex models. This pattern may be driven by researchers' love for their craft, which motivates them to prioritize the full use of available computational resources rather than addressing the policy issue. This translates into, for instance, systematically resorting to Markov chain models despite the availability of representations that are less computationally demanding (Clarke et al. 2005). Additionally, cross-comparing different modelling typologies is a practice that is scarcely explored in the field.

Adopt an 'assumption hunting' attitude: A study's conclusion may depend on an assumption whose impact is untested. In the NCD domain, the most critical assumptions include: the modelling of the dose responses adopted in terms of risk factor estimates; how the dose intake varies upon policy implementation, especially across sociodemographic cohorts and geographical areas; the timeframe of the interventions and their (diminishing/increasing) returns; and the actual NCDs taken into account in the modelling exercise, as well as their change over the timeframe considered (Lo Piano and Robinson 2019).

Detect Garbage In Garbage Out: The strategy of reducing uncertainty in terms of nutrient intake has been adopted to downplay the overall output uncertainty in some modelling activities (Lo Piano and Robinson 2019). Some modellers also produced estimates of the incidence and prevalence of diseases by resorting to educated guesses when information of sufficient quality was not available (European Society of Cardiology 2012).

Find sensitive assumptions before they find you: Uncertainty and sensitivity analyses should be used more widely in the field, especially in light of the number of assumptions made. For instance, the impact of the method used to impute missing data in terms of dose intake on output uncertainty (using other countries' figures or correcting algorithms) could be tested in these settings.

Aim for transparency: Transparent modelling should be the standard for enabling scrutiny from peers *a fortiori* in a decision-making context when models are to be used to inform policies that will eventually influence citizens' lives. For example, the impact of policies to compare the growing prevalence of overweight and obesity in children in the EU was tested in the proprietary model Joint Action on Nutrition and Physical Activity (JANPA), which was not available for public scrutiny (Lo Piano and Robinson 2019).

Do the right sums, not just the sums right: Health evaluations emphasize the economic dimension by glossing over social and cultural aspects of lifestyle and nutrition choice. The latter, however, could be crucial for citizens when prioritizing options. Health and quality of life are captured as per the normative dimensions of citizens' values, which are not necessarily captured through monetary proxies.

Perform thorough, state-of-the-art uncertainty and sensitivity analyses: Sensitivity analysis is known and applied in several nutrition and public health economic evaluations. However, in the vast majority of the cases, these are implemented by varying one factor at a time. As these models are likely to be non-additive, this approach fails to capture interactions between factors. More seriously, in this kind of sensitivity analysis, the vast majority of the output uncertainty space is left unexplored (Saltelli and Annoni 2010). Hence, uncertainty is not adequately characterized or apportioned.

Sociohydrology

The field of sociohydrology has emerged in response to the failure of the traditional, anthropocentric paradigm in water resource management, where human activity is deemed a mere boundary condition of the hydrologic model (Sivapalan, Savenije, and Blöschl 2012a). Sociohydrology seeks to understand how the hydrological cycle changes according to interactions among natural and human forces. Sociohydrology uses Coupled Human and Water Systems (CHAWS) models to address the complex water-related problems currently facing human societies. The key features of CHAWS, which include complexity, cross-scale dynamics, and uncertainty (Sivapalan 2015, Sivapalan, Savenije, and Blöschl 2012b, Wheater and Gober 2013, Liu et al. 2007), make sociohydrology a domain in which the adoption of the PNS paradigm and sensitivity auditing is fully justified.

Check against a rhetorical use of mathematics: CHAWS rest on the assumption that uncertainty reduction and minimization in dynamic and emergent human–water systems can be achieved by adding complexity to the models. This approach is also resorted to in order to rectify the discrepancies between model outputs and observed values (see, for example, Pan et al. (2018)). As a case in point, more parameters are added to justify the social processes included. As things stand in terms of the development of sociohydrology, however, sophistication of the social processes in CHAWS models cannot match the high level of detail seen in hydrologic models due to the lack of knowledge.

Adopt an 'assumption hunting' attitude: CHAWS models contain several assumptions regarding human values, beliefs, and norms related to water use, livelihoods, and the environment (Alonso Vicario et al. 2020, Hemmati et al. 2021). These cultural norms and values drive human behaviour with respect to water resources. They have typically been conceptualized in a 'black box' manner and represented by proxy data (Roobavannan et al. 2018). However, to make a judgement regarding the quality of this parameterization of cultural values, more information would be needed than is currently available in CHAWS.

Detect Garbage In Garbage Out: Risks of future flood damage are underestimated in CHAWS due to issues such as short collective memory, excessive trust in flood protection structures, and a high level of risk-taking in models (Viglione et al. 2014).

Find sensitive assumptions before they find you: The pitfalls of one-factor-at-a-time sensitivity analysis are well known in the hydrology community. However, this approach has been widely used in sociohydrology (see, for example, Liu et al. 2015 and Srinivasan 2015).

Aim for transparency: CHAWS models are not typically open access and are only available for limited case studies. This prevents the community from verifying prior results. In this context, agent-based modelling, for example, is one of the commonly used modelling tools in CHAWS (Shafiee and Zechman 2013, Zhao, Cai, and Wang 2013, Pouladi et al. 2019). However, these models are usually not well documented and lacking in descriptions (Grimm et al. 2006), hence leading to scarcely reproducible results.

Do the right sums, not just the sums right: CHAWS may be attempting to answer the wrong question. It asks 'What is the most likely future?' instead of 'What kind of future do we want and what are the consequences of different policy decisions relative to that desired future?' (Gober and Wheater 2015).

Perform thorough, state-of-the-art uncertainty and sensitivity analyses: In this domain, a systematic global sensitivity analysis can avoid implausible results by exploring and apportioning the output uncertainty to its driving factors and their interactions. It may do this by, for instance, simultaneously considering the uncertainties related to the hydrologic environment (for example, non-stationarity deriving from anthropogenic factors, such as changes in land use, climate, and water use) and the social aspect of CHAWS (socioeconomic development, demography, and agent behaviour) (Sheikholeslami et al. 2019). However, global sensitivity analysis has rarely been performed. Two notable examples are Elshafei et al. (2016) and Ghoreishi, Razavi, and Elshorbagy (2021). These authors investigated the sensitivity of CHAWS to model parameters, which govern the internal dynamics of the system and determine the external sociopolitical context. Ghoreishi et al. (2021) performed a global sensitivity analysis on their agent-based agricultural water demand model to determine the most important model parameters to be calibrated.

Food security

Evaluating the outcomes of humanity's use of the Earth's resources is tightly linked to estimating its overall level of appropriation, particularly in terms of determining the nature of human activities on Earth and the resources needed to sustain them. In this context, a prominent challenge is food security—that is, meeting the nutritional needs of a growing global population, as also recognized in the United Nations agenda for 2030 (United Nations 2021), including a Zero Hunger Strategy cutting across several Sustainable Development Goals. Bahadur KC et al. (2016) proposed their own recipe for achieving healthy and sustainable food provision based on innovative agricultural techniques and dietary re-adaptation. The contribution narrates a successful story from the perspective of a person in 2050 looking back to the past. Saltelli and Lo Piano (2017) used this work as a test benchmark for a deconstruction along the lines of sensitivity auditing.

> *Check against a rhetorical use of mathematics*: The package of policies proposed to assure more sustainable farming and healthy diets consisted of the following: consumer education; increasing the cost of unhealthy food; capturing the environmental costs associated with farming; reducing corn subsidy in the US; and enhancing storage and processing facilities in the developing world. This mixture of policies raises some questions of viability, particularly in relation to potential unintended consequences, as discussed below.
>
> *Adopt an 'assumption hunting' attitude*: The proposed assessment rests on a number of assumptions that are poorly explored. One is neglecting the principle of diminishing return when projecting the yield increase in cultivation to the year 2050—a constant increase in yield over decades hits against known phenomena of topsoil erosion and exhaustion. Additionally, the study assumes a less caloric diet for an increased share of the adult population in the year 2050, and a lower extension of cultivated land globally due to higher yields being obtained with less impactful agricultural techniques. All of these assumptions appear to err on the side of optimism—for example, they do not consider the adjustment costs of the transformation or the realism of popular acceptance of the suggested policies.
>
> *Detect Garbage In Garbage Out*: Reductions in global cultivated land area are estimated at 438 million hectares for the year 2050, with three significant

digits. However, uncertainty in the current yearly global cultivated land extensions amounts to around 20% (around 1000 million hectares according to one of the reference databases (Lo Piano 2017)), making an accuracy of three significant digits for projections to 2050 implausible.

Find sensitive assumptions before they find you: The sensitivity of the output estimates to the input assumptions was left unaddressed in the study published by Bahadur KC et al. (2016).

Aim for transparency: The fact that the model underpinning the quantitative scenario is only available upon request hampers scrutiny by peers and policy makers. Models should be made available along with their quantitative outcomes so as to fully foster their replication and scrutiny.

Do the right sums, not just the sums right: The analysis proposed adopts primarily the standpoint of developed countries in pursuing food security with technical solutions and a policy package tailored to this area of the world. Nevertheless, there remain significant political issues of power asymmetry for developing countries in the context of the international food commodity trade. This translates into an unequal caloric exchange with regard to food crops (Falconí, Ramos-Martin, and Cango 2017) that is not explored in the study of food security. Briefly, a political problem has been reframed into a technical one, while the policy proposals are designed to meet the needs of a minority.

Perform thorough, state-of-the-art uncertainty and sensitivity analyses: The uncertainty space has not been explored, because all the information is conveyed with crisp, uncertainty-free figures.

Conclusion

The case studies examined show how sensitivity analysis can be used in practical terms. These applications open the quantifications to inspection by other disciplines, including social sciences studies. The relation of sensitivity auditing with the loose community of practitioners of PNS is also an element favouring the take-up of its rules, whose spirit—as discussed—is for practical purposes similar to that of the *Manifesto*. The relation between modelling and society has been made more intense and at the same time more conflicted by the COVID-19 pandemic (Pielke 2020, Rhodes and Lancaster 2020). The present crisis of trust in expertise also affects mathematical modelling, and has links to the political crisis affecting several mature democracies under the paradigm that 'solutions to the problem of knowledge are solutions to the problem of social order' (Shapin and Schaffer 2011: 387). Sociology of quantification is actively mapping this territory of conflict for the case of statistics, yet more work is expected for mathematical modelling.

Criticizing the political use of statistics in what he calls '[g]overnance driven quantification', Salais (2010) talks about the reversal of the statistical pyramid, whereby instead of statistics generating concepts and categories useful for collective learning, indicators are produced to demonstrate that preselected policies are efficiently achieved. This is the known nemesis of evidence-based policy into policy-based evidence; among scholars, there is an expectation that these instrumental uses of evidence might come under increasing criticism (van Zwanenberg 2020) (see chapter 7).

References

Alonso Vicario, S., M. Mazzoleni, S. Bhamidipati, M. Gharesifard, E. Ridolfi, C. Pandolfo, and L. Alfonso. 2020. 'Unravelling the Influence of Human Behaviour on Reducing Casualties during Flood Evacuation'. *Hydrological Sciences Journal*, 65(14), 2359–2375, doi: 10.1080/02626667.2020.1810254.

Araujo, L., A. Saltelli, and S.V. Schnepf. 2017. 'Do PISA Data Justify PISA-Based Education Policy?' *International Journal of Comparative Education and Development*, 19(1), 20–34, doi: 10.1108/ IJCED-12-2016–0023.

Bahadur KC, K., E.F. Krishna, G. Dias, T. Zundel, and S. Pascoal. 2016. 'Pathways Leading to a More Sustainable and Healthy Global Food System'. *The Solutions Journal*, 7(5), 10–12.

Borges, J. 1941. *El Jardin de los Senderos que se Bifurcan*, SUR, Buenos Aires.

Breznau, N., E.M. Rinke, A. Wuttke, H.H.V. Nguyen, M. Adem, J. Adriaans, A. Alvarez-Benjumea, H.K. Andersen, D. Auer, F. Azevedo, O. Bahnsen, D. Balzer, G. Bauer, P.C. Bauer, M. Baumann, S. Baute, V. Benoit, J. Bernauer, C. Berning, ... T. Żółtak. 2022. 'Observing Many Researchers Using the Same Data and Hypothesis Reveals a Hidden Universe of Uncertainty', *Proceedings of the National Academy of Sciences*, 119(44), e2203150119, doi: 10.1073/pnas.2203150119.

Bruno, I., E. Didier, and J. Prévieux. 2014. *Statactivisme. Comment lutter avec des nombres*, Zones, La Découverte, Paris.

Christie, M., A. Cliffe, P. Dawid, and S.S. Senn. 2011. *Simplicity, Complexity and Modelling*, Wiley, New York.

Clarke, P.M., A.M. Gray, A. Briggs, R.J. Stevens, D.R. Matthews, R.R. Holman, and on behalf of the UK Prospective Diabetes Study (UKPDS). 2005. 'Cost-Utility Analyses of Intensive Blood Glucose and Tight Blood Pressure Control in Type 2 Diabetes (UKPDS 72)'. *Diabetologia*, 48(5), 868–77, doi: 10.1007/s00125-005-1717-3.

De Leonardis, O. 2022. 'Quantifying Inequality: From Contentious Politics to the Dream of an Indifferent Power'. In *The New Politics of Numbers: Utopia, Evidence and Democracy*, edited by A. Mennicken and R. Salais, Executive Politics and Governance. Springer International Publishing, Cham, pp. 135–66, doi: 10.1007/978-3-030-78201-6_5.

Elshafei, Y., M. Tonts, M. Sivapalan, and M.R. Hipsey. 2016. 'Sensitivity of Emergent Sociohydrologic Dynamics to Internal System Properties and External Sociopolitical Factors: Implications for Water Management'. *Water Resources Research* 52(6), 4944–66, doi: 10.1002/2015WR017944.

Espeland, W.N. and M. Sauder. 2016. *Engines of Anxiety: Academic Rankings, Reputation, and Accountability*, Russell Sage Foundation, New York.

European Society of Cardiology. 2012. *European Cardiovascular Disease Statistics 2012*, Tech. Rep. European Heart Network & European Society of Cardiology, Sophia Antipolis, France.

Falconí, F., J. Ramos-Martin, and P. Cango. 2017. 'Caloric Unequal Exchange in Latin America and the Caribbean'. *Ecological Economics*, *134*, 140–9, doi: 10.1016/j.ecolecon.2017.01.009.

Ferretti, F., A. Saltelli, and S. Tarantola. 2016. 'Trends in Sensitivity Analysis Practice in the Last Decade'. *Science of The Total Environment*, *568*, 666–70, doi: 10.1016/j.scitotenv.2016. 02.133.

Funtowicz, S. and J.R. Ravetz. 1990. *Uncertainty and Quality in Science for Policy*, Kluwer, Dordrecht, doi: 10.1007/978-94-009-0621-1_3.

Funtowicz, S. and J.R. Ravetz. 1993. 'Science for the Post-Normal Age'. *Futures*, *25(7)*, 739–55, doi: 10.1016/0016-3287(93)90022- L.

Gelman, A. and E. Loken. 2013. 'The Garden of Forking Paths: Why Multiple Comparisons Can Be a Problem, Even When There Is No "Fishing Expedition" or "p-Hacking" and the Research Hypothesis Was Posited Ahead of Time', *Statistical Modeling, Causal Inference, and Social Science* [Blog], https://statmodeling. stat.columbia.edu/2021/03/16/the-garden-of-forking-paths-why-multiple-comparisons-can-be-a-problem-even-when-there-is-no-fishing-expedition-or-p-hacking-and-the-research-hypothesis-was-posited-ahead-of-time-2/ (last accessed 14 April 2023).

Ghoreishi, M., S. Razavi, and A. Elshorbagy. 2021. 'Understanding Human Adaptation to Drought: Agent-Based Agricultural Water Demand Modeling in the Bow River Basin, Canada'. *Hydrological Sciences Journal*, *66(97)*, 389–407, doi: 10.1080/02626667.2021.1873344.

Gober, P. and H.S. Wheater. 2015. 'Debates—Perspectives on Socio-Hydrology: Modeling Flood Risk as a Public Policy Problem'. *Water Resources Research*, *51(6)*, 4782–8, doi: 10.1002/2015WR016945.

Grimm, V., U. Berger, F. Bastiansen, S. Eliassen, V. Ginot, J. Giske, J. Goss-Custard, T. Grand, S.K. Heinz, G. Huse, A. Huth, J.U. Jepsen, C. Jørgensen, W.M. Mooij, B. Müller, G. Pe'er, C. Piou, S.F. Railsback, A.M. Robbins, M.M. Robbins, E. Rossmanith, N. Rüger, E. Strand, S. Souissi, R.A. Stillman, R. Vabø, U. Visser, and D.L. DeAngelis. 2006. 'A Standard Protocol for Describing Individual-Based and Agent-Based Models'. *Ecological Modelling*, *198(1)*, 115–26, doi: 10.1016/j.ecolmodel.2006.04.023.

Hemmati, M., H.N. Mahmoud, B.R. Ellingwood, and A.T. Crooks. 2021. 'Unraveling the Complexity of Human Behavior and Urbanization on Community Vulnerability to Floods'. *Scientific Reports*, *11(1)*, 20085, doi: 10.1038/s41598-021-99587-0.

Jasanoff, S. 2003. 'Technologies of Humility: Citizen Participation in Governing Science'. *41(3)*, 223–44, doi: 10.1023/A:1025557512320.

Liu, D., F. Tian, M. Lin, and M. Sivapalan. 2015. 'A Conceptual Socio-Hydrological Model of the Coevolution of Humans and Water: Case Study of the Tarim River Basin, Western China'. *Hydrology and Earth System Sciences*, *19(2)*, 1035–54, doi: https://doi.org/10.5194/hess-19-1035-2015.

Liu, J., S.R.T. Dietz, M.A. Carpenter, C. Folke, E. Moran, A.N. Pell, P. Deadman, T. Kratz, J. Lubchenco, E. Ostrom, Z. Ouyang, W. Provencher, C.L. Redman, S.H. Schneider, and W.W. Taylor. 2007. 'Complexity of Coupled Human and Natural Systems'. *Science 317(5844)*, 1513–16, doi: 10.1126/science.1144004.

Lo Piano, S. 2017. 'Quantitative Story-Telling and Sensitivity-Auditing Appraisal of Expert Solutions to the Issue of Food Security' [Presentation], Conference: Post-Normal Science Symposium PNS3—University of Tuebingen, doi: 10.13140/RG.2.2.36288.43525.

Lo Piano, S. and L. Benini. 2022. 'A Critical Perspective on Uncertainty Appraisal and Sensitivity Analysis in Life Cycle Assessment'. *Journal of Industrial Ecology, 26(3)*, 763–81, doi: 10.1111/jiec.13237.

Lo Piano, S., M. Janos Lorincz, A. Puy, S. Pye, A. Saltelli, S. Thor Smith, and J.P. van der Sluijs. 2022. *Unpacking the Modelling Process for Energy Policy Making* [Preprint], doi: 10.48550/arXiv. 2111.00782.

Lo Piano, S. and M. Robinson. 2019. 'Nutrition and Public Health Economic Evaluations under the Lenses of Post Normal Science'. *Futures, 112*, 102436, doi: 10.1016/J.FUTURES.2019.06.008.

Mennicken, A. and W.N. Espeland. 2019. 'What's New with Numbers? Sociological Approaches to the Study of Quantification'. *Annual Review of Sociology, 45(1)*, 223–45, doi: 10.1146/annurev-soc-073117-041343.

Mennicken, A. and R. Salais, eds. 2022. *The New Politics of Numbers*, Palgrave Macmillan, London.

Micklewright, J. and S.V. Schnepf. 2006. *Inequality of Learning in Industrialised Countries*, 2517, Institute of Labor Economics (IZA), Oxford University Press, Oxford.

Micklewright, J., S.V. Schnepf, and C. Skinner. 2012. 'Non-Response Biases in Surveys of Schoolchildren: The Case of the English Programme for International Student Assessment (PISA) Samples'. *Journal of the Royal Statistical Society: Series A (Statistics in Society), 175(4)*, 915–38, doi: 10.1111/j.1467-985X.2012.01036.x.

Millgram, E. 2015. *The Great Endarkenment: Philosophy for an Age of Hyperspecialization*, Oxford University Press, Oxford.

Pan, H., B. Deal, G. Destouni, Y. Zhang, and Z. Kalantari. 2018. 'Sociohydrology Modeling for Complex Urban Environments in Support of Integrated Land and Water Resource Management Practices'. *Land Degradation and Development, 29(10)*, 3639–52, doi: 10.1002/ldr.3106.

Pielke, R. Jr. 2020. 'The Mudfight Over "Wild-Ass" Covid Numbers Is Pathological'. *Wired*, 22 April 2020 [Preprint]. https://www.wired.com/story/the-mudfight-over-wild-ass-covid-numbers-is-pathological (last accessed 20 March 2023).

Popp Berman, E. and D. Hirschman. 2018. 'The Sociology of Quantification: Where Are We Now?' *Contemporary Sociology, 47(3)*, 257–66.

Pouladi, P., A. Afshar, M.H. Afshar, A. Molajou, and H. Farahmand. 2019. 'Agent-Based Socio-Hydrological Modeling for Restoration of Urmia Lake: Application of Theory of Planned Behavior'. *Journal of Hydrology, 576*, 736–48, doi: 10.1016/j.jhydrol.2019.06.080.

Puy, A., S. Lo Piano, and A. Saltelli. 2020. 'Current Models Underestimate Future Irrigated Areas'. *Geophysical Research Letters, 47(8)*, e2020GL087360, doi: 10.1029/2020GL087360.

Refsgaard, J.C., J.P. van der Sluijs, J. Brown, and P. van der Keur. 2006. 'A Framework for Dealing with Uncertainty Due to Model Structure Error'. *Advances in Water Resources, 29(11)*, 1586–97.

Rhodes, T. and K. Lancaster. 2020. 'Mathematical Models as Public Troubles in COVID-19 Infection Control: Following the Numbers'. *Health Sociology Review*, *29*, 1–18, doi: 10.1080/14461242.2020.1764376.

Roobavannan, M., T.H.M. van Emmerik, Y. Elshafei, J. Kandasamy, M.R. Sanderson, S. Vigneswaran, S. Pande, and M. Sivapalan. 2018. 'Norms and Values in Socio-hydrological Models'. *Hydrology and Earth System Sciences*, *22*(2), 1337–49, doi: 10.5194/hess-22-1337-2018.

Salais, R. 2010. 'La donnée n'est pas un donné. Pour une analyse critique de l'évaluation chiffrée de la performance'. *Revue française d'administration publique*, *135*(3), 497–515, doi: 0.3917/rfap.135.049.

Saltelli, A. 2018. 'Why Science's Crisis Should Not Become a Political Battling Ground'. *Futures*, *104*, 85–90.

Saltelli, A., K. Aleksankina, W. Becker, P. Fennell, F. Ferretti, N. Holst, S. Li, and Q. Wu. 2019. 'Why So Many Published Sensitivity Analyses Are False: A Systematic Review of Sensitivity Analysis Practices'. *Environmental Modelling & Software*, *114*, 29–39, doi: 10.1016/J.ENVSOFT.2019.01.012.

Saltelli, A. and P. Annoni. 2010. 'How to Avoid a Perfunctory Sensitivity Analysis'. *Environmental Modelling & Software*, *25*(12), 1508–17, doi: 10.1016/j.envsoft.2010.04.012.

Saltelli, A., G. Bammer, I. Bruno, E. Charters, M. Di Fiore, E. Didier, W.N. Espeland, J. Kay, S. Lo Piano, D. Mayo, R. Pielke Jr, T. Portaluri, T.M. Porter, A. Puy, I. Rafols, J.R. Ravetz, E. Reinert, D. Sarewitz, P.B. Stark, A. Stirling, J. van der Sluijs, and P. Vineis. 2020a. 'Five Ways to Ensure That Models Serve Society: A Manifesto'. *Nature*, *582*(7813), 482–4, doi: 10.1038/d41586-020-01812-9.

Saltelli, A., L. Benini, S. Funtowicz, M. Giampietro, M. Kaiser, E.S. Reinert, and J.P. van der Sluijs. 2020b. 'The Technique Is Never Neutral. How Methodological Choices Condition the Generation of Narratives for Sustainability'. *Environmental Science and Policy*, *106*, 87–98.

Saltelli, A. and B. D'Hombres. 2010. 'Sensitivity Analysis Didn't Help. A Practitioner's Critique of the Stern Review'. *Global Environmental Change*, *20*(2), 298–302, doi: 10.1016/j.gloenvcha.2009.12. 003.

Saltelli, A., D.J. Dankel, M. Di Fiore, N. Holland, and M. Pigeon. 2022. 'Science, the Endless Frontier of Regulatory Capture'. *Futures*, *135*, 102860, doi: 10.1016/j.futures.2021.102860.

Saltelli, A. and S. Funtowicz. 2014. 'When All Models Are Wrong'. *Issues in Science and Technology*, *30*(2), 79–85.

Saltelli, A. and S. Lo Piano. 2017. 'Problematic Quantifications: A Critical Appraisal of Scenario Making for a Global "Sustainable" Food Production'. *Food Ethics*, *1*(2), 173–9.

SAPEA. 2018. 'SAPEA Participates in Scientific Integrity Conference', *SAPEA*, 20 February 2018, https://sapea.info/sapea-participates-in-scientific-integrity-conference/ (last accessed 14 April 2023).

Scoones, I. and A. Stirling. 2020. *The Politics of Uncertainty*, edited by I. Scoones and A. Stirling, Routledge, Abingdon, New York, doi: 10.4324/9781003023845.

Shafiee, E. and E.Z. Berglund. 2013. 'An Agent-Based Modeling Framework for Sociotechnical Simulation of Water Distribution Contamination Events'. *Journal of Hydroinformatics*, *15*(3), 862–80, doi: 10.2166/hydro.2013.158.

Shapin, S. and S. Schaffer. 2011. *Leviathan and the Air-Pump: Hobbes, Boyle, and the Experimental Life*, Princeton Classics, Princeton University Press, Princeton, NJ.

Sheikholeslami, R., S. Razavi, H.V. Gupta, W. Becker, and A. Haghnegahdar. 2019. 'Global Sensitivity Analysis for High-Dimensional Problems: How to Objectively Group Factors and Measure Robustness and Convergence While Reducing Computational Cost'. *Environmental Modelling & Software*, 111, 282–99, doi: 10.1016/j.envsoft.2018.09.002.

Sivapalan, M. 2015. 'Debates—Perspectives on Socio-Hydrology: Changing Water Systems and the "Tyranny of Small Problems"—Socio-Hydrology'. *Water Resources Research*, 51(6), 4795–805, doi: 10.1002/2015WR017080.

Sivapalan, M., H.H.G. Savenije, and G. Blöschl. 2012a. 'Socio-Hydrology: A New Science of People and Water'. *Hydrological Processes*, 26(8), 1270–6, doi: 10.1002/hyp.8426.

Sivapalan, M., H.H.G. Savenije, and G. Blöschl. 2012b. 'Socio-Hydrology: A New Science of People and Water'. *Hydrological Processes*, 26(8), 1270–6, doi: 10.1002/hyp.8426.

Srinivasan, V. 2015. 'Reimagining the Past – Use of Counterfactual Trajectories in Socio-Hydrological Modelling: The Case of Chennai, India'. *Hydrology and Earth System Sciences*, 19(2), 785–801, doi: 10.5194/hess-19-785-2015.

Thévenot, L. 2022. 'A New Calculable Global World in the Making: Governing through Transnational Certification Standards'. In *The New Politics of Numbers: Utopia, Evidence and Democracy*, edited by A. Mennicken and R. Salais, Executive Politics and Governance, Springer International Publishing, Cham, pp. 197–252, doi: 10.1007/978-3-030-78201-6_7.

United Nations. 2021. 'Transforming Our World: The 2030 Agenda for Sustainable Development', Department of Economic and Social Affairs. https://sdgs.un.org/2030agenda (last accessed 14 April 2023).

van der Sluijs, J.P., M. Craye, S. Funtowicz, P. Kloprogge, J.R. Ravetz, and J. Risbey. May 2005. 'Combining Quantitative and Qualitative Measures of Uncertainty in Model-Based Environmental Assessment: The NUSAP System'. *Risk Analysis*, 25(2), 481–92, doi: 10.1111/j.1539-6924.2005.00604.x.

van Zwanenberg, P. 2020. 'The Unravelling of Technocratic Orthodoxy'. In *The Politics of Uncertainty*, edited by I. Scoones and A. Stirling, Routledge, Abingdon, pp. 58–72.

Viglione, A., G. Di Baldassarre, L. Brandimarte, L. Kuil, G. Carr, J. Luis Salinas, A. Scolobig, and G. Blöschl. 2014. 'Insights from Socio-Hydrology Modelling on Dealing with Flood risk—Roles of Collective Memory, Risk-Taking Attitude and Trust'. *Journal of Hydrology*, 518, 71–82, doi: 10.1016/j. jhydrol.2014.01.018.

Wheater, H. and P. Gober. 2013. 'Water Security in the Canadian Prairies: Science and Management Challenges'. *Philosophical Transactions of the Royal Society A: Mathematical, Physical and Engineering Sciences*, 371, 20120409, doi: 10.1098/rsta.2012.0409.

Woessmann, L. 2016. 'The Economic Case for Education'. *Education Economics*, 24(1), 3–32, doi:10.1080/09645292.2015.1059801.

Wuttke, J. 2007. *Uncertainty and Bias in PISA*, SSRN Scholarly Paper ID 1159042, Social Science Research Network, Rochester, NY.

Zhao, J., X. Cai, and Z. Wang. 2013. 'Comparing Administered and Market-Based Water Allocation Systems through a Consistent Agent-Based Modeling Framework'. *Journal of Environmental Management*, 123, 120–30, doi: 10.1016/j.jenvman.2013.03.005.

9

Mathematical modelling

Lessons from composite indicators

Marta Kuc-Czarnecka and Andrea Saltelli

Introduction

As early as the beginning of the seventeenth century, Galileo turned his attention to the quantification of immeasurable phenomena: count what is countable, measure what is measurable, and what is not measurable, make measurable. Fast forward to our times, and the 'make measurable' injunction is today visible in practices that would have surprised Galileo, including in the profusion of composite indicators (CIs) that are the subject of the present chapter.

The ambition to capture complex phenomena for analysis and advocacy has resulted in an explosion of metrics. Combinations of many diagnostic variables are now routinely taken as the basis for the ordering and (or) grouping of objects (people, regions, countries, enterprises). Although the cult of quantification is nothing new (Mau 2019), the 'measure-mania' (Diefenbach 2009) has grown significantly with the expansion of computerization and digitization, leading to what has been termed the 'tyranny of metrics' (Muller 2018), the situation whereby more and more aspects of life come to be characterized by numbers (Mennicken and Espeland 2019).

We have become accustomed to these measures, often accepting them without concerning ourselves with the way they are constructed. Countries are ranked according to happiness, the degree of their energy transition, academic freedom, the rule of law, and so on, over a myriad of domains. Cities, universities, and even individuals are ranked, though in the latter case the operation is performed within systems we normally have no access to, such as in the case of credit scores produced by proprietary algorithms, and used by banks to offer or deny a mortgage. These latter would not be counted as CIs, although they often apply similar logics, This is because most CIs have communication as a primary aim, and hence, in general, are quite explicit in revealing details of their construction (Stiglitz, Sen, and Fitoussi 2009).

Hence, CIs have become part of the extended and growing family of numbers that are so much part of our existence. We barely reflect on them: 'too close to

Marta Kuc-Czarnecka and Andrea Saltelli, *Mathematical modelling*. In: *The politics of modelling*. Edited by: Andrea Saltelli and Monica Di Fiore, Oxford University Press. © Oxford University Press (2023).
DOI: 10.1093/oso/9780198872412.003.0009

us, they have become part of the very lens through which we attend to and comprehend the world' (Saltelli et al. 2021). Numerification of reality, via both visible and invisible numbers, has also exploded as an active field of academic investigation in several disciplines: in sociology of quantification (Espeland and Stevens 2008, Popp Berman and Hirschman 2018, Mennicken and Espeland 2019, Mennicken and Salais 2022), but also in law and economics, and among practitioners of quantification themselves, from data analysts to statisticians. This literature also highlights crucial indications as to the hidden consequences of quantification—as discussed in the introduction to this volume—in the movement of statactivists (Bruno, Didier, and Prévieux 2014, Samuel 2022), campaigns against image recognition (Algorithmic Justice League 2020), and in other initiatives.

But what is a CI? For Michael Power (2004), CIs belong to the family of second-order measurement (meta-measurement)—that is, numbers based on numbers. For Power, these measures take on a life of their own,

> in many cases constituting distinct bodies of knowledge in which averages, measures of dispersion and correlations enable organizations, countries and regions to be compared with each other and for inferential models to be constructed. [...] The objects of second-order measurement must be thought of, and measured, as real and independent of measuring processes.

Thus they obey the suggestion of sociologist and statistician Alain Desrosières (1998) to 'treat social facts as things, after a motto initially proposed by Émile Durkheim'.

According to the Organisation for Economic Co-operation and Development (OECD 2008), CIs are mathematical combinations (or aggregations) of a set of indicators that are used to summarize complex or multidimensional issues. They find use in placing countries' performance at the centre of the policy arena, or in offering a rounded assessment of countries' performance, as well as in ranking a variety of other entities, from cities to universities. Whether perceived as a threat or as an opportunity, these measures are here to stay, and the OECD (2008) offers a useful list of pros and cons (Box 9.1).

Box 9.1 Pros and cons of CIs

Pros:

- Allow the inclusion of more information within existing size limits.
- Place issues of country performance and progress at the centre of the policy arena.
- Facilitate communication with the general public (i.e. citizens, media, etc.) and promote accountability.
- Help to construct/underpin narratives for lay and literate audiences.
- Enable users to compare complex dimensions effectively.

Cons:

- May disguise serious failings in some dimensions and increase the difficulty of identifying proper remedial action if the construction process is not transparent.
- May lead to inappropriate policies if dimensions of performance that are difficult to measure are ignored.

Source: Adapted from OECD 2008.

Looking today at this table of pros and cons, the list of cons—while short—appears to describe well the many sins attributed to CIs by more recent works (O'Neil 2016, Muller 2018, Mennicken and Salais 2022). Today, the political dimensions of rating and ranking appear to have taken centre-stage. Without denying the use of these measures for well-meaning advocacy, for example for fiscal justice, many academicians point to their utilization in the context of New Public Management (NPM) theories (Diefenbach 2009) at the service of some form of 'neoliberal' ideology, the political doctrine that revived the concept of the supremacy of markets and advocated for small governments in the 1980s (see also 'The public sector' in chapter 6). Here CIs would serve to evaluate performance and reward profit and efficiency (Crouch 2011) while ignoring other important dimensions such as fairness and equity. For example, according to the universal measures of university appeal based on international ranking, important elements of the historical mission of universities disappear (Mittelman 2017).

As discussed below, Kuc (2017) shows how a country's domestic divergence is often the hidden price of inter-country convergence. Salais (2022) discusses how metrics of employment are likewise built in a way that hides the precarity and insecurity of jobs in order to maximize the rate of employment.

The campaign 'Doing Rights Not Rankings', signed by over 360 signatories from 80 countries, called the World Bank and its shareholders to end the Ease of Doing Business rankings (Ghosh 2020, Mariotti 2021). The World Bank Doing Business project was blamed for inviting countries to adopt neoliberal policies and reducing social programmes. Staying with these examples (employment, convergence, doing business), the problem is that a country achieving a high score does not translate into substantial improvement in the situation of people excluded from the labour market, living in socially and economically backward regions, or falling out of the safety net of social programmes. Another example is the CIs of energy transition (World Economic Forum (WEF) 2021) and their focus on renewable green energy sources. Here the 'rich north' grants itself the moral right to ban the 'poor south' from using fossil fuels. As noted by Stiglitz, Sen, and Fitoussi (2009), the main problem with CIs, it is not the lack of information about their technical make-up, but the fact that the normative assumptions on which the measure is based are often left unspoken. Porter (2012) calls the logic behind these measures that of 'thin prescriptions', an allusion to the depth that remains excluded from sight. For Power (2004), these second-order measurements participate in dynamic cycles where trust and distrust alternate, as old measures are replaced by new ones following institutional and political dynamics—all of which we are seeing, at the time of writing, in the process of 'refreshing' the logic of Ease of Doing Business with a brand-new index (Cobham 2022).

There exists a rich critical literature related to the construction process of CIs (for example Berger and Bristow 2009, Saisana, Saltelli, and Tarantola 2005, Paruolo, Saisana, and Saltelli 2013, Becker et al. 2017, Gnaldi and Del Sarto 2018, Cartone and Postiglione 2020, Cinelli et al. 2020, Kuc-Czarnecka, Lo Piano, and Saltelli 2020). The OECD *Handbook on Constructing Composite Indicators* (OECD 2008) has existed for over a decade. Nevertheless, not all CIs appear to follow existing recommended practices. For example, analyses of the uncertainty and sensitivity (see chapter 5) are rare, which is an obstacle to their transparency. Mainstream economists offer solutions based on 'normalizing' CI by expressing all dimensions in dollars thanks to shadow prices. For World Bank's economist Martin Ravallion (2010), composite indices should be built based on economic theory, using direct monetary aggregates as in Gross Domestic Product (GDP), or based on shadow prices. Other indicators, such as the Human Development Index (HDI) and the Multidimensional Poverty Index (MPI), are labelled by Ravallion as 'mashup indices'. In our opinion, reducing a CI to a cost–benefit analysis begs the question of whether and when this latter makes sense (see also 'The numbers critique' in chapter 6).

Still, new CIs are born at an increasing pace (Bandura 2011, Yang 2014, Greco et al. 2019; see Fig. 9.1), and all actors use them, from governments to international organizations to non-governmental organizations (NGOs). CIs, when used as instruments of advocacy, lend themselves to political controversy and conflict: we tend to approve of those whose normative premises we share, and dislike the others. Those who oppose the World Bank's Doing Business indicator may like the work of the Tax Justice Network (2022) for fiscal equity, or that of the Electoral Integrity Project (Carrie 2017) concerned with upholding democracy. When a multidimensional phenomenon is locked in (and looked at) from a single perspective, chances are someone will disagree.

For observers who adhere to NPM theories, numbers are facts remote from values. 'Often, immersion in the facts makes value disagreements feel much less relevant', notes Cass Sunstein (Matthews 2018). For those of a more philosophical disposition, the theory of Amartya Sen on the informational basis for judgement in justice (IBJJ) offers a way to look at measures such as CI in two ways: in terms of technical adequacy and of fairness. For the former, quality is concerned with the data and their aggregation. Fairness corresponds instead to whether the measure permits a fair, informed judgement of an issue. In Sen's capabilities framework, fairness implies that a measure takes into consideration the possibility of

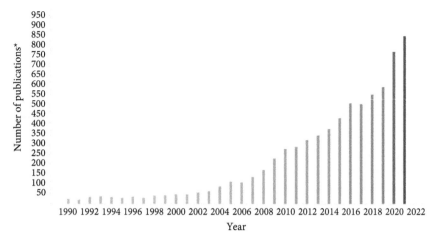

Fig. 9.1 Scopus query, searching for the strings 'composite indicator*' and 'composite index' in the title, abstract of keywords of a paper.

each individual, depending on their status, achieving the life they aspire to; fairness is not arrived at through equal material means but through having equal opportunities to fulfil one's aspirations (Sen 1990).

In this chapter, we discuss CIs as an instance of mathematical modelling and demonstrate how CIs can be improved using the postulates contained in the *Manifesto* (Saltelli et al. 2020). To this effect, we discuss aspects of each of the postulates using existing CIs as examples.

Mind the assumptions: Assess uncertainty and sensitivity

There are undoubtedly many uncertainties, modelling choices, and normative implications in the construction of CIs. An error propagation applied to a CI may show that the volatility of a ranking is considerable (Saisana, Saltelli, and Tarantola 2005, Saisana and Saltelli 2010). Even assuming reliable theoretical foundations and appropriately selected variables, construction stages such as normalization, aggregation, and the weighting of variables play a crucial role in determining the index in any application field.

The process of weighing variables introduces an obvious element of normative selection in the measure, as noted in (Stiglitz, Sen, and Fitoussi 2009). The expedient of resorting to not weighting, for example, corresponds to the tacit assumption that all variables have equal status (European Commission 2020, in the section 'Weighting'). The application of sensitivity analysis to CI in many cases also reveals inconsistencies in the way weights are interpreted as measures of importance (Paruolo, Saisana, and Saltelli 2013, Becker et al. 2017, Lindén et al. 2021a).

These last authors propose their CIAO tool (The Composite Indicator Analysis and Optimization Matlab Toolbox). This toolbox allows users to determine to what extent a given variable explains the final ranking and to adjust the weights so that the influence of a given variable is consistent with a predefined assumption, among other things. In applying this (Lindén et al. 2021b), it may turn out that the weights assigned by the creators of CIs do not reflect the individual factors' actual degree of influence. They may rather reflect the developers' understanding of the importance, which is not always confirmed by experience; for example, a scatterplot of the index versus a variable may reveal the variable to be much more or less important than the developer anticipated.

In fact, it is easy to confuse weights and importance: imagine, for example, a dean who builds a homemade CI to rank the professors in her/his departments using hours spent teaching and the number of papers written, without weighting. This results in professors being ranked based mostly on the papers. In this stylized example, this may happen because the variables have not been standardized, and the number of papers varies more than the hours spent teaching. The dean then increases the weight of the hours of teaching, so that both variables concur in the ranking. When the professors look at the work of the dean, they complain that 'publishing does not count in this department', as they note the higher weight given to the teaching. They are wrong, of course. The weight is not the same as the importance.[1]

A problem that often arises is the spurious 'importance' given to certain variable because of their correlation with other diagnostic variables, while the given variable alone does not carry additional information. Despite the recommendations of (OECD 2008), these correlations are not always taken into account or are ignored during the CI's construction stage (Cinelli et al. 2020). In a study undertaken by *Times Higher Education* (THE) on the ranking of higher education (HE), the two variables based on the opinions of prospective recruiters and academics are much more important than stipulated by the THE developers, with the consequence that THE is more opinion-based and less fact-based than its developers would have it (Paruolo, Saisana, and Saltelli 2013). In this as in other cases, highly correlated variables lead to a double-counting that biases the final index. An additional problem appears when the synthetic measure includes negatively correlated variables, pointing to a logical inconsistency in the measure (Saisana and Philippas 2012).

Paruolo, Saisana, and Saltelli (2013) and Lindén et al. (2021a) discuss ways to improve the assignment of weights in CI built via linear aggregation. Due to its intuitive appeal, linear aggregation is often chosen in the construction of indices. Since time immemorial, polymaths and philosophers have struggled against the linear combination of decision variables; Ramon Llull (1232–1315) and Nicholas of Cusa (1401–64) developed methods that were rediscovered

---

[1] The example is discussed in full by Paruolo, Saisana, and Saltelli (2013).

more recently by Nicolas de Condorcet (1743–94) and Jean-Charles de Borda (1733–99) respectively (Munda 2008). The reasons why linear aggregations—while intuitive—are methodologically poor are many, including the confusion between weights and importance already mentioned. While we cannot enter into this here, we can mention the important problem of compensation. When adding variables, a good score for one variable can compensate for a poor score in another area. Most citizens would prefer a country with good GDP and healthy infrastructure to one with excellent GDP and poor infrastructure. Additionally, adding variables implies that they are independent of the viewpoint of voter preferences, which is often not the case (see 'Procrustes' quantifauxcation' in chapter 2).

In conclusion, we recommend that the developer—when building a CI or reassessing it for a given year—engage in an analysis of the robustness of a CI with methods derived from sensitivity analysis. We can refer to this analysis as a 'modelling of the modelling process', in the sense that the developers try to check systematically the implications of all steps taken in the analysis, from normalization all the way to aggregation. This will allow internally consistent measures to be obtained, as well as in some cases reducing the set of diagnostic variables. Additionally, where there are reasons to suspect that the weighting has introduced inconsistencies, the approaches suggested in Paruolo, Saisana, and Saltelli (2013) and Lindén et al. (2021a) could be considered.

Mind the hubris: Complexity can be the enemy of relevance

The known asymmetry between model developers—who are aware of the limitations of their construct—and their users—who are in general less so—can also be at play for CIs, so that more is read into rankings than the aggregation's quality would support. A CI may make use of a large number of input variables, which convey an impression of complexity and completeness. When evidence shows that the ranking depends only upon a smaller subset of variables, one can suspect a rhetorical use of numericized evidence. Several important CIs developed by renowned institutions, such as the Global Competitiveness Index (GCI), Energy Transition Index (ETI), or Digital Economic and Society Index (DESI), may fall into this category (Olczyk, Kuc-Czarnecka, and Saltelli 2022; Kuc-Czarnecka, Olczyk, and Zinecker 2021; Olczyk and Kuc-Czarnecka 2022). CIs based on a wide set of diagnostic variables may suffer from opacity. When many variables contribute to the index, the effect of individual variables is blurred, which is undesirable in policy terms because of their relevance in real life. Olczyk, Kuc-Czarnecka, and Saltelli (2022) showed that in the case of GCI 4.0, 35 out of 103 variables are practically 'silent', as the information they transfer to the final index is close to zero. Most of the variables that were insignificant in explaining the GCI variance turned out to be opinion-based indicators—another consequence of the internal correlation among variables discussed in the previous section. Similar results hold for DESI: a very similar linear ordering of countries results from halving the set of diagnostic variables (from 37 to 18) and removing one of the pillars entirely (Olczyk and Kuc-Czarnecka 2022). As is the case for models, complexity may reduce the relevance of CI, especially when using linear aggregations of variables. However, one should also be aware of the political economy of quantification (Power 2004; see also 'The political economy of mathematical modelling' in chapter 13), whereby complexity increases the epistemic authority of the measurement.

Mind the framing: Match purpose and context

As discussed above, the objective of a CI is often that of advocacy (Saltelli 2007). Thus, CIs strongly depend upon a selected normative frame that might be accepted by some constituencies and rejected by others. Adopting multiple frames could be a solution (Kuc-Czarnecka, Lo Piano, and Saltelli 2020). As noted by Salais (2022), 'No data, especially aggregated data, can be said to be the truth, not only because they are deeply linked to the series of both technical and political choices made along their chain of production, but basically because among a range of possible choices, one path only has been chosen' (p. 385).

This is true of many CIs that consider there to be 'only one correct' perspective. Kuc-Czarnecka, Lo Piano, and Saltelli (2020) tested the concept of the CI at variable geometry, meaning that different structures and dimensions were explored to make allowances for different visions or 'storytelling' of an issue. This was applied to indicators of cohesion policy in the EU. The analysis assumed four possible ways of looking at cohesion, ideally corresponding to different stakeholders. The first viewpoint takes variables related to the labour market and social protection as its inputs. Another viewpoint corresponds to a stakeholder who asks for measures of governance and fairness to be included. A third stakeholder prefers measures of healthcare to those of governance, while a fourth stakeholder wants them both. The results show that countries tend to converge if the first two options are chosen, and to diverge if instead viewpoints three and four are adopted. Thus, countries differ more when aspects such as accountability and stability, corruption, political functioning, regulatory quality, and the rule of law are considered. Kuc-Czarnecka, Lo Piano, and Saltelli (2020) suggest modifying the philosophy of CIs from 'analysis cum advocacy' to create 'analysis with multiple storytelling'.

Clearly, this approach would be more appealing in a context of negotiation than in one of pure advocacy. When advancing a particular storyline, one does not wish to be encumbered by alternative or opposing visions. Instead, negotiation may call for a multitude of scenarios, actors, and points of view to be considered. Transparency in the frames puts the use of CIs into a context of social discovery (Dewey 1927, Boulanger 2014), allowing different visions to be contested and compared, as suggested by the French statactivists (Bruno, Didier, and Prévieux 2014). Ensuring that the final ranking is not the subjective creation of its architect may be a hard goal to achieve, but a discussion around possible architectures can ensure that while valuable statistical information feeds into the decision, it does not construct too partial an account of the situation. Exposing the recipient of rankings and groupings to their 'double' activates a process of social discovery and promotes a more informed, less technocratic approach to the decisions in matters of complex phenomena. Statactivists in particular note how the viewpoint of those 'measured'—the knowledge they possess without even knowing it—may contribute to the fairness of the analysis (Salais 2022).

Mind the consequences: Quantification may backfire

A growing multidisciplinary literature has engaged with the consequences of the diffusion of metrics for evaluation and classification. How could quantification backfire? A system of indicators can trigger a process of goal displacement—the so-called Goodhart's law, whereby actors adapt to the measure and 'game it' (Muller 2018). In other words, actors are diverted into maximizing metrics, sometimes at the expense of the real objectives. In response to Goodhart's effect, a logical 'rule cascade' response may be triggered, whereby new rules are introduced to counteract the gaming. These rules, in turn, trigger a new cycle of gaming. According to Muller, other unwanted or unexpected consequences of excess rating include discouraging practices that might be fruitful in themselves, such as risk-taking, innovation, cooperation, and common purpose.

Among the most impactful examples of this 'audit culture' (Strathern 2000) is the case of university rankings, which have led to important changes in the nature of HE (Lee 2008), including the almost paradoxical result that HE administrations may now recruit a 'ranking strategist'[2] in order to improve their university's score. Most volumes written to give warning about the misuse of ratings and rankings mention the ranking of HE as an egregious case of a measure that backfired (Muller 2018, O'Neil 2016), while specific works exist to address the harm produced by these measures (Mittelman 2017, Espeland and Sauder 2016). Welsh (2021) points out that these rankings 'are necessarily productive and reproductive of oligarchy', meaning that political power is something inherent in rankings, so they are not neutral and passive, but contribute to the creation of a 'regime of institutional control'. As a remedy, he suggests that we should 'make a break in

---

[2] 'The University of Saskatchewan is hiring a University Ranking strategist'. Source: https://twitter.com/LizzieGadd/status/1456750554072469520 (last accessed 1 April 2023).

particular from the positivism, institutionalism, and managerialism that characterizes prevailing approaches' which are a manifestation of the neoliberal political economy. Welsh calls the rankings 'crypto-feudal technology of control'.

Lee (2008) notes that universities are overconcerned with moving up ranking tables at the expense of the quality of education, as these rankings do not 'reward' teaching. In his opinion, focusing on a specific measure negatively affects the quality of the education process, making it not just unhealthy but fatally ill. For Mittelman, while rankings are a concause rather than a cause of the present 'implausible dream' of globalized world universities, rankings have contributed to the erosion of the three 'historic/foundational' missions of HE: cultivating democratic citizenship, fostering critical thinking, and protecting academic freedom. Rankings have accelerated a redefinition of the objectives of HE towards efficiency and excellence, sent prices rocketing up for students and their families, increased inequality, and increased the power of HE administrators relative to that of their academic colleagues (Mittleman 2017).

Remaining in the domain of education, accusations of perverse consequences have also targeted the Programme for International Students Assessment (PISA). Since the publication of its first results in 2000, the rank has been a subject of controversy, charged with irrelevance and counter-productivity (Zhao 2020) (see also 'The Programme for International Student Assessment (PISA)' in chapter 8). Hopfenbeck et al. (2018) lament the lack of representativeness and inadequate sampling. Solheim and Lundetræ (2018) point to linguistic and cultural bias, while Zhao (2020) focuses on PISA's methodological flows, for example the way in which the Rasch model is used. The critique of Araujo, Saltelli, and Schnepf (2017) registers PISA's normative stance, foremost the fact that education is investigated as an input to growth—to the point that some investigators link PISA score points to GDP points. Hopmann (2008) notes that there is no empirical confirmation that PISA measures key competences and skills necessary in the modern world. According to him, PISA's overwhelming success is the result of illusion and self-marketing, in which Western European societies have chosen to believe.

As mentioned above, the best-known reaction to ranking is the aforementioned 'Doing Rights Not Rankings' campaign, the main purpose of which was to suspend the publication of the *Doing Business* report. The authors of the campaign pointed out that the World Bank, using its influence, contributes to the deterioration of working conditions, the increase in inequality, the erosion of workers' rights, and the growing gender and racial gap (AFRODAD 2021). Mariotti (2021) studied what he calls data fabrication, especially in relation to some countries. For example, the positions of Azerbaijan, China, Saudi Arabia, and United Arab Emirates were lower than expected because of irregularities. This author assumes that the irregularities are not involuntary, but the result of a precise design. According to Ghosh (2020), the ranking is the result of the disproportionate power of the US and people having too much faith in the pro-cyclical nature of economies.

Mind the unknowns: Acknowledge ignorance

Often, and especially in the use of evidence-based policy, a political problem is transformed via quantification into a technical problem (Ravetz 1971). In these settings, one moves from evidence-based policy to policy-based evidence. This is where technocratic approaches to governance (van Zwanenberg 2020) combine reductionism—whereby complex phenomena reduce to a linear metric—with a suppression of ignorance (Scoones and Stirling 2020). In the case of CIs, their virtue of reifying complex dimensions (Box 9.1) is thus also their sin, as the capture is achieved by transforming the qualitative (for example a concept such as rule of law or fiscal fairness) into a quantitative, with all the associated risks of reductionism, justificationism, and purported neutrality. Acknowledging ignorance, in the case of CIs, implies moving from advocacy and analysis towards negotiation, in the space left open by the consideration that more possible frames and futures are possible. One example of such a negotiation on different ways to measure poverty is offered by the French statactivists (Concialdi 2014), and one on consumer prices indices by Samuel (2022).

Conclusion

As noted in the *Manifesto* for models, CIs may help to frame more questions than answers. In this respect, the concept of extended peer community might be useful, alongside the theory of post-normal science (PNS; Funtowicz and Ravetz 1993, 1994; see also 'Involving stakeholders' in chapter 3 and 'Healing the amputation of awareness in science' in chapter 12). An extended peer community forms when experts from different disciplines merge with supposed non-experts, such as the lay public, investigative journalists, whistleblowers, and whoever has a stake or an interest in the matter being debated. In PNS, such a community should be deliberative rather than purely consultative and should not refrain from using data or statistical measures, provided these come with their pedigrees. Pedigrees, introduced by Funtowicz and Ravetz (1990) (see also 'The epistemic background of sensitivity auditing' in chapter 8), ensure something similar to Sen's required IBJJ—that is, a chance for the various publics to form an informed opinion, and an opinion that includes an understanding of the ideological stances and stakes of the various actors, including of the proponents of a measure. In this respect, one positive characteristic of CIs is that by attracting attention to an issue (Box 9.1), they may by themselves catalyse the creation of an extended peer community, albeit one of antagonists, as was the case with the Doing Business Index discussed here. One can only hope that the academic world might one day muster the necessary energy to oppose international rankings of HE, if the authors quoted in this chapter are right that these are so damaging to the existence of universities and democracy.

As with the example of statactivists—whose motto is 'fight a number with a number'—a growing social criticism towards rankings produced by powerful institutions appears to be a social process that facilitates a pluralism of 'data-making'. For CIs, one should also keep in mind Muller's (2018) warning that, considering the possible shortcomings, 'the best use of metrics *may be* not to use it at all' (our emphasis).

References

AFRODAD. 2021. 'Doing Business Report Campaign: Doing Rights Not Rankings'. *AFRODAD*. https://afrodad.org/doing-business-report-campaign-doing-rights-not-rankings (last accessed 1 April 2023).

Algorithmic Justice League. 2020. 'Algorithmic Justice League: Unmasking AI Harms and Biases'. https://www.ajl.org/ (last accessed 28 April 2022).

Araujo, L., A. Saltelli, and S.V. Schnepf. 2017. 'Do PISA Data Justify PISA-Based Education Policy?' *International Journal of Comparative Education and Development*, *19(1)*, 20–34, doi: 10.1108/IJCED-12-2016-0023.

Bandura, R. 2011. 'Composite Indicators and Rankings: Inventory 2011. Technical Report', Office of Development Studies, United Nations Development Programme (UNDP), New York.

Becker, W., M. Saisana, P. Paruolo, and I. Vandecasteele. 2017. 'Weights and Importance in Composite Indicators: Closing the Gap'. *Ecological Indicators*, *80*, 12–22, doi: 10.1016/j.ecolind.2017.03.056.

Berger, T. and G. Bristow. 2009. 'Competitiveness and the Benchmarking of Nations—A Critical Reflection'. International *Advances in* Economic Research, *15*, 378–92, doi: 10.1007/s11294-009-9231-x.

Boulanger, P.-M. 2014. 'Elements for a Comprehensive Assessment of Public Indicators', Report EUR 26921EN. http://publications.jrc.ec.europa.eu/repository/bitstream/JRC92162/lbna26921enn.pdf (last accessed 1 April 2023).

Bruno, I., E. Didier, and J. Prévieux. 2014. *Statactivisme. Comment lutter avec des nombres*, Zones, La Découverte, Paris.

Carrie, A. 2017. 'The Mathematicians Who Want to Save Democracy'. *Nature*, *546(7657)*: 200–2, doi: 10.1038/546200a.

Cartone, A. and P. Postiglione. 2020. 'Principal Component Analysis for Geographical Data: the Role of Spatial Effects in the Definition of Composite Indicators'. *Spatial Economic Analysis*, *16(2)*, 126–47, doi: 10.1080/17421772.2020.1775876.

Cinelli, M., M. Kadziński, M. Gonzalez, and R. Słowiński. 2020. 'How to Support the Application of Multiple Criteria Decision Analysis? Let Us Start with a Comprehensive Taxonomy'. *Omega*, *96*, 102261, doi: 10.1016/j.omega.2020.102261.

Cinelli, M., M. Spada, W. Kim, Y. Zhang, and P. Burgherr. 2020. 'MCDA Index Tool: an Interactive Software to Develop Indices and Rankings'. *Environment Systems and Decisions*, *41*, 81–109, doi: 10.1007/s10669-020-09784-x.

Cobham, A. 2022. 'What the BEEP? The World Bank Is Doing Business Again'. *Tax Justice Network*, 17 March 2022. https://taxjustice.net/2022/03/17/what-the-beep-the-world-bank-is-doing-business-again (last accessed 1 April 2023).

Concialdi, P. 2014. 'Le BIP40: Alerte Sur La Pauvreté'. In *Statactivisme. Comment Lutter Avec Des Nombres*, edited by I. Bruno, E. Didier, and J. Prévieux, Zones, La Découverte, Paris, pp. 199–211.

Crouch, C. 2011. *The Strange Non-Death of Neo-Liberalism*, Polity Press, Cambridge.

Desrosières, A. 1998. *The Politics of Large Numbers: A History of Statistical Reasoning*, Harvard University Press, Cambridge, MA.

Dewey, J. 1927. *The Public and Its Problems. An Essay in Political Inquiry*, Swallow Press, Athens, OH.

Diefenbach, T. 2009. 'New Public Management in Public Sector Organizations: The Dark Side of Managerialistic "Enlightenment"'. *Public Administration, 87*(4), 892–909, doi: 10.1111/j.1467-9299.2009.01766.x.

Espeland, W.N. and M. Sauder. 2016. *Engines of Anxiety: Academic Rankings, Reputation, and Accountability*, Russell Sage Foundation, New York.

Espeland, W.N., and M.L. Stevens. 2008. 'A Sociology of Quantification'. *European Journal of Sociology, 49*(3), 401–36, doi: 10.1017/S0003975609000150.

Funtowicz, S., and J.R. Ravetz. 1990. *Uncertainty and Quality in Science for Policy*, Kluwer, Dordrecht, doi: 10.1007/978-94-009-0621-1_3.

Funtowicz, S., and J.R. Ravetz. 1993. 'Science for the Post-Normal Age'. *Futures, 25*(7), 739–55, doi: 10.1016/0016-3287(93)90022-L.

Funtowicz, S., and J.R. Ravetz. 1994. 'The Worth of a Songbird: Ecological Economics as a Post-Normal Science'. *Ecological Economics, 10*(3), 197–207, doi: 10.1016/0921-8009(94)90108-2.

Ghosh, J. 2020. '"Stop Doing Business": Project Syndicate'. *Project Syndicate*, 10 September 2020. https://www.project-syndicate.org/commentary/world-bank-should-scrap-doing-business-index-by-jayati-ghosh-2020-09?barrier=accesspaylog (last accessed 1 April 2023).

Gnaldi, M. and S. Del Sarto. 2018. 'Variable Weighting via Multidimensional IRT Models in Composite Indicators Construction'. *Social Indicators Research, 136*, 1139–56, doi: 10.1007/s11205-016-1500-5.

Greco, S., A. Ishizaka, M. Tasiou, and G. Torrisi. 2019. 'On the Methodological Framework of Composite Indices: A Review of the Issues of Weighting, Aggregation, and Robustness'. *Social Indicators Research, 141*, 61–94, doi: 10.1007/s11205-017-1832-9.

Hopfenbeck, T.N., J. Lenkeit, Y. El Masri, K. Cantrell, J. Ryan, and J.-A. Baird. 2018. 'Lessons Learned from PISA: A Systematic Review of Peer-Reviewed Articles on the Programme for International Student Assessment'. *Scandinavian Journal of Educational Research, 62*(3), 333–53, doi: 10.1080/00313831.2016.1258726.

Hopmann, S.T. 2008. 'No Child, No School, No State Left Behind: Schooling in the Age of Accountability'. *Journal of Curriculum Studies, 40*(4), 417–56, doi: 10.1080/00220270801989818

European Commission. 2020. 'Your 10-Step Pocket Guide to Composite Indicators & Scoreboards, Competence Centre on Composite Indicators and Scoreboards'. https://knowledge4policy.ec.europa.eu/publication/your-10-step-pocket-guide-composite-indicators-scoreboards_en (last accessed 14 April 2023).

Kuc, M. 2017. 'Is the Regional Divergence a Price for the International Convergence? The Case of the Visegrad Group', *Journal of Competitiveness, 9*(4), 50–65, doi: 10.7441/joc.2017.04.04.

Kuc-Czarnecka, M., S. Lo Piano, and A. Saltelli. 2020. 'Quantitative Storytelling in the Making of a Composite Indicator'. *Social Indicators Research*, *149*, 775–802, doi: 10.1007/s11205-020-02276-0.

Kuc-Czarnecka, M.E., M. Olczyk, and M. Zinecker. 2021. 'Improvements and Spatial Dependencies in Energy Transition Measures'. *Energies*, *14(13)*, 3802, doi: 10.3390/en14133802.

Lee, H. 2008. 'Rankings of Higher Education Institutions: A Critical Review'. *Quality in Higher Education*, *14(3)*, 187–207, doi: 10.1080/13538320802507711.

Lindén, D., M. Cinelli, M. Spada, W. Becker, P. Gasser, and P. Burgherr. 2021a. 'A Framework Based on Statistical Analysis and Stakeholders' Preferences to Inform Weighting in Composite Indicators'. *Environmental Modelling and Software*, *145*, 105208, doi: 10.1016/j.envsoft.2021.105208.

Lindén, D., M. Cinelli, M. Spada, W. Becker, and P. Burgherr. 2021b. *Composite Indicator Analysis and Optimisation (CIAO) Tool, v.2*, doi: 10.13140/RG.2.2.14408. 75520.

Mariotti, C. 2021. 'How Many Scandals Will It Take for the World Bank to Start Doing Rights Not Rankings?' *European Network on Debt and Development*, 18 March 2021, https://www.eurodad.org/how_many_scandals_will_it_take_for_the_world_bank_to_start_doing_rights_not_rankings (last accessed 1 April 2023).

Matthews, D. 2018. 'Can Technocracy Be Saved? An Interview with Cass Sunstein'. *Vox*, 22 October 2018, https://www.vox.com/future-perfect/2018/10/22/18001014/cass-sunstein-cost-benefit-analysis-technocracy-liberalism (last accessed 1 April 2023).

Mau, S. 2019. *The Metric Society: On the Quantification of the Social*, Polity Press, Cambridge.

Mennicken, A., and W.N. Espeland. 2019. 'What's New with Numbers? Sociological Approaches to the Study of Quantification'. *Annual Review of Sociology*, *45(1)*, 223–45, doi: 10.1146/annurev-soc-073117-041343.

Mennicken, A., and R. Salais, eds. 2022. *The New Politics of Numbers: Utopia, Evidence and Democracy*, Palgrave Macmillan, London.

Mittelman, J.H. 2017. *Implausible Dream: The World-Class University and Repurposing Higher Education*, Princeton University Press, Princeton, NJ.

Muller, J.Z. 2018. *The Tyranny of Metrics*, Princeton University Press, Princeton, NJ.

Munda, G. 2008. *Social Multi-Criteria Evaluation for a Sustainable Economy*, Springer, Berlin.

OECD. 2008. *Handbook on Constructing Composite Indicators: Methodology and User Guide*, OECD Publishing, Paris.

Olczyk, M., and M. Kuc-Czarnecka. 2022. 'Digital Transformation and Economic Growth: DESI Improvement and Implementation'. *Technological and Economic Development of Economy*, *28*, 775–803, doi: 10.3846/tede.2022.16766.

Olczyk, M., M. Kuc-Czarnecka, and A. Saltelli. 2022. 'Changes in the Global Competitiveness Index 4.0 Methodology: The Improved Approach of Competitiveness Benchmarking'. *Journal of Competitiveness*, *14(1)*, 118–35, doi: 10.7441/joc.2022.01.07.

O'Neil, C. 2016. *Weapons of Math Destruction: How Big Data Increases Inequality and Threatens Democracy*, Random House Publishing Group, Portland, OR.

Paruolo, P., M. Saisana, and A. Saltelli. 2013. 'Rating and Rankings: Voodoo or Science?' *Journal of Royal Statistical Society Series A: Statistics in Society*, *3*, 609–34, doi: 10.1111/j.1467-985X.2012.01059.x.

Popp Berman, E., and D. Hirschman. 2018. 'The Sociology of Quantification: Where Are We Now?' *Contemporary Sociology*, 47(3), 257–66, doi: 10.1177/0094306118767649.

Porter, T.M. 2012. 'Funny Numbers'. *Culture Unbound Journal of Current Cultural Research*, 4(4), 585–98, doi: 10.3384/cu.2000.1525.124585.

Power, M. 2004. 'Counting, Control and Calculation: Reflections on Measuring and Management'. *Human Relations*, 57(6), 765–83, doi: 10.1177/0018726704044955.

Ravallion, M. 2010. 'Mashup Indices of Development', Policy Research Working Paper Series, The World Bank, 5432, doi: 10.1596/1813-9450-5432.

Ravetz, J.R. 1971. *Scientific Knowledge and Its Social Problems*, Transaction Publishers, Piscataway Township, NJ.

Saisana, M. and D. Philippas. 2012. *Sustainable Society Index (SSI): Taking Societies' Pulse along Social, Environmental and Economic Issues*, OCLC: 847462966, European Commission, Joint Research Centre, Institute for the Protection and the Security of the Citizen, Publications Office, Luxembourg.

Saisana, M., A. Saltelli, and S. Tarantola. 2005. 'Uncertainty and Sensitivity Analysis as Tools for the Quality Assessment of Composite Indicators'. *Journal of the Royal Statistical Society*, 168(2), 307–23, doi: 10.1111/j.1467-985X.2005.00350.x.

Saisana, M. and A. Saltelli. 2010. *Uncertainty and Sensitivity Analysis of the 2010 Environmental Performance Index*, https://publications.jrc.ec.europa.eu/repository/handle/JRC56990 (last accessed 14 April 2023).

Salais, R. 2022. 'La Donnée n'est Pas Un Donné': Statistics, Quantification and Democratic Choice'. In *The New Politics of Numbers: Utopia, Evidence and Democracy*, edited by A. Mennicken, and R. Salais, Executive Policy and Governance, Palgrave Macmillan, Cham, pp. 379–415.

Saltelli, A. 2007. 'Composite Indicators Between Analysis and Advocacy'. *Social Indicators Research*, 81, 65–77, doi: 10.1007/s11205-006-0024-9.

Saltelli, A., A. Andreoni, W. Drechsler, J. Ghosh, R. Kattel, I.H. Kvangraven, I. Rafols, E.S. Reinert, A. Stirling, and T. Xu. 2021. 'Why Ethics of Quantification Is Needed Now', UCL Institute for Innovation and Public Purpose, Working Paper Series (IIPP WP 2021/05). https://www.ucl.ac.uk/bartlett/public-purpose/publications/2021/jan/why-ethics-quantification-needed-now (last accessed 1 April 2023).

Saltelli, A. and M. Di Fiore. 2020. 'From Sociology of Quantification to Ethics of Quantification'. *Humanities and Social Sciences Communications*, 7(69), doi: 10.1057/s41599-020-00557-0.

Saltelli, A., G. Bammer, I. Bruno, E. Charters, M. Di Fiore, E. Didier, W. Nelson Espeland, J. Kay, S. Lo Piano, D. Mayo, R. Pielke Jr, T. Portaluri, T.M. Porter, A. Puy, I. Rafols, J.R. Ravetz, E. Reinert, D. Sarewitz, P.B. Stark, A. Stirling, J. van der Sluijs, and P. Vineis. 2020. 'Five Ways to Ensure That Models Serve Society: A Manifesto'. *Nature*, 582(7813), 482–4, doi: 10.1038/d41586-020-01812-9. PMID: 32581374.

Samuel, B. 2022. 'The Shifting Legitimacies of Price Measurements: Official Statistics and the Quantification of Pwofitasyon in the 2009 Social Struggle in Guadeloupe'. In *The New Politics of Numbers: Utopia, Evidence and Democracy*, edited by A. Mennicken and R. Salais, Executive Policy and Governance, Palgrave Macmillan, Cham, pp. 337–77.

Scoones, I., and A. Stirling. 2020. 'Uncertainty and the Politics of Transformation'. In *The Politics of Uncertainty Challenges of Transformation*, 1st edition, edited by I. Scoones and A. Stirling, Routledge, London, pp. 1–30.

Sen, A. 1990. 'Justice: Means versus Freedoms'. *Philosophy & Public Affairs*, *19*(2), 111–21.

Solheim, O.J. and K. Lundetræ. 2018. 'Can Test Construction Account for Varying Gender Differences in International Reading Achievement Tests of Children, Adolescents and Young Adults? A Study Based on Nordic Results in PIRLS, PISA and PIAAC'. *Assessment in Education: Principles, Policy and Practice*, *25*(1), 107–26, doi: 10.1080/0969594X.2016.1239612.

Stiglitz, J.E., A. Sen, and J.-P. Fitoussi. 2009. *Report by the Commission on the Measurement of Economic Performance and Social Progress*, OECD. https://www.insee.fr/en/statistiques/fichier/2662494/stiglitz-rapport-anglais.pdf (last accessed 14 April 2023).

Strathern, M. 2000. 'Introduction: New Accountabilities'. In *Audit Cultures*, edited by M. Strathern, Routledge, London, pp. 1–18.

Tax Justice Network. 2022. 'Let's Take Back Control of Our Tax Systems', *Tax Justice Network*, https://taxjustice.net/ (last accessed 1 April 2023).

van Zwanenberg, P. 2020. 'The Unravelling of Technocratic Orthodoxy'. In *The Politics of Uncertainty: Challenges of Transformation*, edited by I. Scoones and A. Stirling, pp. 58–72. Routledge, London.

Welsh, J. 2021. 'A Power-Critique of Academic Rankings: Beyond Managers, Institutions, and Positivism'. *Power and Education*, *13*(1), doi: 10.1177/1757743820986173.

World Economic Forum (WEF). 2021. *Fostering Effective Energy Transition 2021 Edition*, World Economic Forum, Geneva.

Yang, L. 2014. *An Inventory of Composite Measures of Human Progress*, Technical Report, UNDP Human Development Report Office.

Zhao, Y. 2020. 'Two Decades of Havoc: A Synthesis of Criticism Against PISA' . *Journal of Education Change*, *21*, 245–66, doi: 10.1007/s10833-019-09367-x.

10

Mathematical modelling, rule-making, and the COVID-19 pandemic

Ting Xu

Introduction

When in 2020 we adapted our private and professional agendas to the pace of the spread of COVID-19, we all recognized a main lesson of contemporary society: we must learn to live with an increase in risks and uncertainties.[1] The novel coronavirus, SARS-COV-2—about which we knew very little when it emerged—quickly spread all over the world. 'Part of the considerable difficulty in managing this epidemic', a UK government health adviser wrote to Richard Horton, editor-in-chief of *The Lancet*, is that we 'have some major gaps in knowledge (especially around asymptomatic transmission by age)' (Horton 2020: 935). The scale of uncertainty and ignorance about COVID-19 is unnerving (Harford 2020). This has urged rule makers to resort to mathematical modelling when making rules to prevent and contain the spread of the epidemic.

Mathematical modelling has produced various sets of data on COVID-19, including the number of infections and fatalities. Countless governmental rules on preventing the spread of the epidemic, including maintaining social distance, isolation, and lockdowns, are based on these data. The Imperial College team's model updated in March 2020, for example, played a crucial role in changing the UK government's policy on the pandemic (Adam 2020a: 318).

Managing the COVID-19 pandemic is just the latest example of the fundamental challenge to the ways in which rules and decisions are made and to what rule-making can achieve. In an uncertain world, such as one paralysed by a global pandemic, rules need to be made under circumstances where 'facts are uncertain, values in dispute, stakes high and decisions urgent' (Funtowicz and Ravetz 1993: 744). In such a world, rule-making is being increasingly influenced by knowledge

[1] This chapter is based on a longer version published in *Amicus Curiae* (2021). The author would like to thank Roger Cotterrell, Monica Di Fiore, Tim Murphy, Amanda Perry-Kessaris, Andrea Saltelli, Carl Stychin, and Maurice Sunkin for their suggestions and comments on an earlier draft of this chapter. All remaining errors are my responsibility.

Ting Xu, *Mathematical modelling, rule-making, and the COVID-19 pandemic*. In: *The politics of modelling*. Edited by: Andrea Saltelli and Monica Di Fiore, Oxford University Press. © Oxford University Press (2023).
DOI: 10.1093/oso/9780198872412.003.0010

produced by experts across disciplines[2]—the most pertinent example being the reliance upon mathematical models by rule makers. To what extent can rule makers navigate and tame uncertainty through mathematical modelling? Or is their reliance on mathematical modelling increasing the uncertainty of rules?

The COVID-19 pandemic brought to light the need to rethink the interaction between mathematical modelling and rule-making under conditions of uncertainty. However, we do not have a proper taxonomy for understanding and assessing the degree of uncertainty in general, let alone the uncertainty associated with the making of mathematical models more specifically. Outside the relatively circumscribed community of sociologists studying the influence of quantification on governance and the intersection of numbers and power, a serious reflection on using modelling as a governance technique is still lacking. Yet, if these problems remain unsolved, trust in public bodies and the accountability of modellers and rule makers will be undermined.

This aspect is so critical and urgent that a group of scholars have signed a *Manifesto* setting out 'five ways to ensure that models are at the service of the society'. Among the recommendations, Saltelli and his colleagues (2020) tell us to 'pay attention to the unknowns', in line with what philosopher Nicholas of Cusa called the 'docta ignorantia'.

'Pay attention to the unknowns' forces us to examine three interrelated questions: what is the nature of mathematical modelling? What kinds of uncertainty are there in the contested arena of science for rule-making? What ideas and approaches can help us better address the interaction between mathematical modelling and rule-making under conditions of uncertainty?

The nature of mathematical modelling

Models can serve many purposes: they function as one of the critical instruments in many disciplines, including computer science and economics, mediating between theories and the real world. They are used for the examination and elaboration of theories; for the exploration of the processes and consequences of applying theories; for the measurement of risks; for giving precise answers in numbers; and for the justification of intervention measures.

Since the 1980s, there has been a dramatic increase in the use of quantification in governing social life (Rottenburg, Merry, and Park 2015), with numbers and quantification exerting a strong influence on law and governance.[3]

Merry (2016: 1), for example, argued that 'quantification is seductive'. By quantification, she meant 'the use of numbers to describe social phenomena in

---

[2] See, e.g., Murphy 1997.
[3] See, e.g., Perry-Kessaris 2011a, 2011b, 2017, Nelken and Siems 2021.

countable and commensurable terms' (Merry 2016: 1); its seductiveness lies in its capacity to produce numbers which provide 'knowledge of a complex and murky world' (Merry 2016: 1). These numbers, then, formed a basis for rule-making which is also seen as 'scientific' and 'evidence-based' (Merry 2016: 4).[4]

Supiot (2017) argued that human action has been increasingly governed by numbers and quantification rather than by law. Indeed, modelling has become a governance technique and has been heavily relied upon by rule makers to navigate and tame the increasing uncertainty of social and economic life. Assumptions and outputs of models and policy recommendations by modellers have been embedded in law, regulation, and policy.

Johns (2020) pointed out that 'public attention should be devoted to the world-making effects of models'. Models are 'artifacts with politics', establishing and normalizing certain patterns of power and authority (Winner 1980: 134).

This scientific authority generated by mathematical modelling contributed significantly to creating a public consciousness of the need to control an urgent problem, be it health, environmental, geopolitical, or economic.

However, 'the promise of control and prediction rooted in the Cartesian dream of rigorous technical models and precise scientific metrics in handling the uncertainties did not survive the test of a radically uncertain world' (Reinert et al. 2021, 8; see also Scoones and Stirling 2020). Why?

There are many elements that explain the misaligned mediation 'between the theories and the real world' that we have mentioned.

Models are by no means neutral: they are shaped by the modellers' disciplinary orientations and made in a particular context, and they embody the modellers' interests, assumptions, and biases (Saltelli et al. 2020).

Mathematical modelling describes 'our beliefs about how the world functions', using mathematical concepts and languages (Lawson and Marion 2008: 1). 'Different contexts—different markets, social settings, countries, time periods and so on—require different models' (Rodrik 2015: 11).[5] For example, using models that were negligent in attending to uncertainty fuelled the financial crisis (Reinert 2009)[6] and delayed action on COVID-19.

Inaccurate assumptions can be made because of the poor quality of data. Decisions made based on the outputs of models may be misleading. This was evidenced in the early stages of the COVID-19 pandemic, when the detection of numbers of infections, deaths, and tests was limited and reporting was delayed (Jewell, Lewnard, and Jewell 2020: 1893).

Consequently, these numbers could be flawed, misused, or misleading. For example, a prestigious 2019 Global Health Security Index ranked the United States

---

[4] For a critique of 'evidence-based' policy making, see also, e.g., Funtowicz and Saltelli 2015.

[5] For a discussion of studying economics while ignoring the context, see, e.g., McCloskey 2002.

[6] For critiques of how financiers respond to uncertainty around the possibility of a financial crisis, see, e.g., Taleb 2010.

as the safest place to be in case of a pandemic (Johns Hopkins Center for Health Security 2019). Prediction based on these numbers clearly contrasts with what happened in the COVID-19 pandemic.

Mathematical models are imbued with uncertainty. This implies significant limitations for the use of models, especially under circumstances where modelling is based on a paucity of data and a significant degree of abstraction and uncertainty is involved (Spiegelhalter 2019): to predict COVID-19 transmission rates, models rely on hundreds of parameters (Adam 2020b: 533) which are poorly understood (Holmdahl and Buckee 2020). Mathematical models can only 'estimate the relative effect of various interventions in reducing disease burden rather than to produce precise quantitative predictions about extent or duration of disease burdens' (Jewell, Lewnard, and Jewell 2020: 1893; see also Whitty 2015: 4). Nevertheless, 'consumers' of mathematical models, including rule makers, the media, and the public, 'often focus on the quantitative predictions of infections and mortality estimates' (Jewell, Lewnard, and Jewell 2020: 1893).

Mathematical models, enabled by computer simulations, have also been used by rule makers to communicate with the public and establish rules on various interventions (Adam 2020a: 316), while the resulting model uncertainty is not always properly communicated to the public (Holmdahl and Buckee 2020). Instead of justifying the certainty of the rules, estimates from mathematical models about COVID-19 can lead to uncertainty and anxiety, for example when these models estimate tens of thousands of future deaths (Jewell, Lewnard, and Jewell 2020: 1893; see also Rhodes and Lancaster 2020).

Developing a taxonomy of uncertainty

As already mentioned, we know that uncertainty is associated with the making of mathematical modelling and that decisions based on the outputs of models are often made under conditions of uncertainty in crises. It is necessary to dissect this typology by developing a taxonomy of uncertainty to ascertain the nature and degree of uncertainty in modelling and rule-making in crises. Uncertainty in the broad sense means that 'given current knowledge, there are multiple possible future states of nature' (Stewart 2000: 41). It is also defined as 'the result of our incomplete knowledge of the world, or about the connection between our present actions and their future outcomes' (Kay and King 2020: 13). The degree of the connection between current human actions and their future consequences varies, giving rise to different types of uncertainty. But how can we gauge the extent of their connection?

Since World War II, the application of modern probability theory has arisen in several major fields where humans interact with complex technologies, including economics, management science, computer science, and artificial intelligence

(Smithson 1989: 3). The popularity of modern probability theory constitutes a major response to the increasing complexity of technologies and organizations, and provides an alternative to deterministic, mechanical approaches to dealing with the complexity and uncertainty embedded in these systems (Smithson 1989: 3).

Probability theory has been applied to solving legal problems. In fact, 'Leibniz' [Gottfried Wilhelm (von) Leibniz, 1646–1716] early formulations of probability theory were motivated by problems of legal inference' (Smithson 1989: 24). However, there are challenges to applying probability theory in the legal field. Judges and juries make decisions based on evidence which is primarily qualitative rather than quantitative and cannot be easily measured by probability. Judges deal with imperfect information, unreliable witnesses, and doubtful 'facts' (Smithson 1989: 23). Their judgments may also be influenced by socioeconomic and political factors. The law is subject to interpretations and revisions. A misunderstanding of probability in judgments may even turn on the 'prosecutor's fallacy': a fallacy in the use of statistical reasoning to test an occurrence. Prominent examples include the rape conviction of Andrew Deen in 1990 based on DNA evidence and the heart-breaking case of Sally Clark in 1999 (Chivers 2021).[7]

The study of uncertainty deals with probability and other concepts, such as vagueness, which cannot be explained in probabilistic language. To decipher the complexity of uncertainty under which rules are made to manage the COVID-19 pandemic, it is important to evaluate different kinds of uncertainty. This can be divided into at least two types according to the degree of connection between current human actions and their future consequences, namely external risk (resolvable uncertainty) and manufactured risk (radical uncertainty, with ignorance as one important dimension).[8] In developing this taxonomy of uncertainty, it is important to bear in mind that we cannot 'apply probabilities to every instance of our imperfect knowledge of the future' (Kay and King 2020: 12).

Analysis of risk and the risk society figure prominently in the writings of some preeminent sociologists researching modernity, in particular Ulrich Beck and Anthony Giddens. For example, Beck's definition of risk, which differs from dangers, is closely associated with reflexive modernization. He argued that risk may be defined as '*a systematic way of dealing with hazards and insecurities induced and*

---

[7] In 1990, Andrew Deen was convicted of rape. His conviction was based partly on DNA evidence and a statement from an expert witness that 'the chance that the DNA came from someone else was just one in 3m' (Chivers 2021). In 1999, Sally Clark was convicted of murdering her two children. Her conviction was again based partly on an expert witness statement that 'the chance of two babies dying of sudden infant death syndrome (Sids) in one family was one in 73m'. Her conviction was overturned in 2003 for not taking into account 'the prior probability—that is, the likelihood that someone was a double murderer, which is, mercifully, even rarer than Sids' (Chivers 2021).

[8] For the difference between external and manufactured risk, see Giddens 1999. For the difference between resolvable uncertainty and radical uncertainty, see Kay and King 2020.

introduced by modernization itself (Beck 1992: 21, italics original). By 'reflexive modernization', Beck meant that modernization 'is becoming its own theme':

> Questions of the development and employment of technologies (in the realms of nature, society and the personality) are being eclipsed by questions of the political and economic 'management' of the risks of actually or potentially utilized technologies—discovering, administering, acknowledging, avoiding or concealing such hazards with respect to specially defined horizons of relevance.
>
> (Beck 1992: 19–20)

Giddens (1999: 3) also pointed out that risk differs from hazard or danger, but he emphasized that 'the idea of risk is bound up with the aspiration to control and particularly with the idea of controlling the future'. The term 'risk society' was coined in the 1980s and became a popular topic in the 1990s. Risk society refers to 'a society increasingly preoccupied with the future (and also with safety), which generates the notion of risk' (Giddens and Pierson 1998, 209). The reason why a risk society is preoccupied with the future, as Giddens further explained in his Chorley Lecture,[9] is that 'we increasingly live on a high technological frontier which absolutely no one completely understands and which generates a diversity of possible futures' (Giddens 1999: 3).

The analysis of uncertainty and risk so far has shown that these two concepts are closely related, as both are the result of our incomplete knowledge of the world and its connections with possible futures. That said, there are differences between risk and uncertainty. Economists (used to) highlight the distinction between risk and uncertainty: risk refers to 'unknowns which could be described with probabilities', while uncertainty means unknowns which could not be described with probabilities (Kay and King 2020: 12). Frank Knight (1885–1972) and John Maynard Keynes (1883–1946) were the two key figures in economics who argued for the continued importance of the distinction. Knight (1921: 20), for example, argued that risk is 'measurable', and that we should 'restrict the term "uncertainty" to cases of the non-quantitative type'. Keynes (1937, 214) pointed out that for uncertainty 'there is no scientific basis on which to form any calculable probability whatever'. Keynes (1937: 222) also argued that 'the hypothesis of a calculable future leads to [...] an underestimation of the concealed factors of utter doubt, precariousness, hope and fear'. These factors are ubiquitous in rule-making in managing crises such as the COVID-19 pandemic but are easily concealed by the illusion of truth and control generated by mathematical modelling.[10] This is despite flaws in the theoretical underpinnings and the quality and quantity of data, as well as biases and human fallibility in making mathematical models. The distinction

---

[9] This lecture was delivered at the London School of Economics on 27 May 1998.

[10] See, e.g., Drechsler (2011: 50, arguing that it is wrong to assume that the use of mathematics necessarily leads to 'truth').

between risk and uncertainty has been sidelined by mainstream economics over the last century through applying 'probabilities to every instance of our imperfect knowledge of the future' (Kay and King 2020: 12).

Given the differences and overlaps between risk and uncertainty, attempts have been made to specify various categories of risk and uncertainty to make sense of the distinctions and reveal the overlaps. Giddens (1999: 4) made a distinction between 'external and manufactured risk'. External risk is 'risk of events that may strike individuals unexpectedly (from the outside, as it were) but that happen regularly enough and often enough in a whole population of people to be broadly predictable, and so insurable' (Giddens 1999: 4). Manufactured risk refers to 'new risk environments for which history provides us with very little previous experience. We often don't really know what the risks are, let alone how to calculate them accurately in terms of probability tables' (Giddens 1999: 4). The COVID-19 pandemic, for example, provides a new risk environment in which we can rely little on experience with previous epidemics. It is a noteworthy example of manufactured risk. Kay and King have chosen to replace the distinction between risk and uncertainty with a distinction between 'resolvable and radical uncertainty':

> Resolvable uncertainty is uncertainty which can be removed by looking something up (I am uncertain which city is the capital of Pennsylvania) or which can be represented by a known probability distribution of outcomes (the spin of a roulette wheel). With radical uncertainty, however, there is no similar means of resolving the uncertainty—we simply do not know. Radical uncertainty has many dimensions: obscurity; ignorance; vagueness; ambiguity; ill-defined problems; and a lack of information that in some cases but not all we might hope to rectify at a future date.
>
> (Kay and King 2020: 14)

The distinction between external and manufactured risk made by Giddens, and the distinction between resolvable and radical uncertainty drawn by Kay and King, help us develop a taxonomy of uncertainty. External risk has overlaps with resolvable uncertainty, while manufactured risk comes close to radical uncertainty. Most challenges to rule-making in managing the COVID-19 pandemic come from manufactured risk or radical uncertainty. Further, radical uncertainty cannot be described in terms of well-defined, numerical probabilities (Kay and King 2020). Increasing radical uncertainty has led to more complexity in rule-making. Scoones and Stirling (2020: 12), for example, argued that claims to be able to control uncertainty seem to 'underpin the securing of authority, justification, legitimacy, trust and wider public acceptance'. But for both modellers and rule makers, if they simply make these claims without awareness of the nature of risk and uncertainty, and the importance of communicating 'the unknown' to the public, their accountability will be undermined.

Re-evaluating the interaction between mathematical modelling and rule-making: Concepts and approaches

Important dimensions of radical uncertainty that must be examined are ignorance and the ways in which modellers and rule makers should deal with it. Understanding the notion and function of 'ignorance' and the dynamism between ignorance and knowledge in the context of rule-making is of primary importance. Under conditions of radical uncertainty, ignorance is unavoidable; however, it is often regarded as 'either the absence or the distortion of "true" knowledge' (Smithson 1989: 1). The conventional approach to ignorance is therefore to eliminate or tame it by using some kind of 'scientific method' (Smithson 1989: 1). The ways in which ignorance has been tamed by mathematical modelling through probabilistic judgements in the context of rule-making in the COVID-19 pandemic is a noteworthy example of this conventional approach being exercised. The key problem in managing the COVID-19 pandemic, however—to echo the observation of Friedman (2005: xiv) regarding the nature of ignorance—is not just gaps in our knowledge about the virus. Rather, the central problem is that the information presented by experts to rule makers, even if supported by mathematical models, only provides a veneer of certainty and may mislead.

Ignorance is 'socially constructed and negotiated' and 'multiple' (Smithson 1989: 6). Acknowledging this reminds us of the famous quote from Rumsfeld on 'known unknowns' and 'unknown unknowns': 'There are known unknowns. That is to say there are things that we now know we don't know. But there are also unknown unknowns. There are things we don't know we don't know.'[11] There are other instances of ignorance which have been overlooked by Rumsfeld, such as 'what we don't know we know' (Rayner 2012). Accordingly, there are various ways that modellers and rule makers try to cope with ignorance, such as taming it with a technical solution. This leads to the use of mathematical models as tools of displacement: 'an object or activity, such as a computer model, designed to inform management of a real-world phenomenon actually becomes the object of management' (Rayner 2012: 120).

Taming ignorance in mathematical modelling and rule-making leads to several consequences. Modellers and rule makers fail to understand that uncertainties are 'conditions of knowledge itself' (Scoones and Stirling 2020: 4). The ways in which we 'understand, frame and construct possible futures' are 'hard-wired into "objective" situations' (ibid.: 4), including the application of probabilities to every kind of uncertainty. Modellers may tend to use models to justify predetermined agendas and to undermine the importance of communicating what is not known

---

[11] Rayner 2012: 107–8, quoting Former US Secretary for Defence, Donald Rumsfeld, NATO Headquarters, 6 June 2002.

(Saltelli et al. 2020). They may also offload accountability to the models they chose (Saltelli et al. 2020).[12] Thus, in managing the COVID-19 pandemic, the UK government's claim that it is 'following the science' has been criticized by scientists as a way to 'abdicate responsibility for political decisions' (Devlin and Boseley 2020).

In some disciplines, including philosophy, sociology, and economics, there have been challenges to the conventional way we approach the nature and function of ignorance. After all, 'learned ignorance' (*docta ignorantia*), or self-awareness of ignorance, was regarded as a virtue for most of the history of Western philosophy (Ravetz 2015). Popper (1985: 55) argued that:

> the more we learn about the world, and the deeper our learning, the more conscious, specific, and articulate will be our knowledge of what we do not know, our knowledge of our ignorance. For this, indeed, is the main source of our ignorance—the fact that our knowledge can only be finite, while our ignorance must necessarily be infinite.

Hayek (1945), as well as Knight and Keynes discussed above, questioned the nature of 'perfect knowledge' and warned us of the dangers of projecting excessive certainty about the future (Davies and McGoey 2012: 76). However, the virtue of learned ignorance has been neglected in contemporary society. Rather, it is a common assumption that modern society is based on the accumulation of reliable and calculable knowledge. Yet, past crises such as the 2007–8 financial crisis have taught us important lessons. In contrast to the common assumption, many institutions that survived the financial crisis are not those which had a firm faith in the reliability of credit-rating agencies. Rather, they were those 'most able to suggest risks were unknowable or not predictable in advance' (Davies and McGoey 2012: 65). The financial crisis thus was not only an economic crisis but also an epistemological and scientific one (Davies and McGoey 2012: 66; see also Best 2010).

Ignorance and knowledge are not antithetical, as Nietzsche argued (2003 [1973]: 24).[13] Rather, ignorance should be regarded as the 'refinement' of knowledge. Ignorance can serve as 'a productive force' and 'the twin and not the opposite of knowledge' (McCoey 2012). This does not mean that there is more value in knowing what we don't know as opposed to not knowing what we don't know. 'Knowing what we don't know' is productive in the sense that it generates a constant need for solutions to crises that experts and rule makers failed to identify earlier (Davies and McCoey 2012: 79). We need to 'lean into the reality of not

---

[12] See also Snow 2021.
[13] See also Ravetz 1987: 100.

knowing' (Snow 2021), and even embrace uncertainty (Scoones and Stirling 2020: 11; see also Ravetz 2015):

> In embracing uncertainty in modelling practice, the emphasis must therefore shift towards active advocacy of qualities of doubt (rather than certainty), scepticism (rather than credulity) and dissent (rather than conformity)—and so towards creative care rather than calculative control. With indeterminacy thus embraced and irreducible plurality accepted, non-control and ignorance emerge as positive values in any attempt to create narratives for policy under conditions of uncertainty.

Conclusion

A great deal of mathematical modelling and rule-making during crises is about coping with uncertainty. For modellers and rule makers, this poses a fundamental challenge, as there has been a lack of a rigorous framework through which to understand and analyse the nature and function of uncertainty in the context of rule-making. Although in some disciplines, including philosophy, sociology, and economics, there have been new studies of and reflection on the way we approach uncertainty and ignorance, responses from legal studies have been slow. Rule makers rely heavily on numbers and quantification, trying to give precise answers and assert control when making decisions in crises. The interdisciplinary literature on uncertainty and ignorance clearly shows us the need to examine the interaction between mathematical modelling and rule-making. This requires a three-step analysis.

First, examine the nature of mathematical modelling, which is closely associated with uncertainty. Mathematical models are also by no means neutral: they are shaped by the modellers' disciplinary orientations, made in a particular context, and embody the modellers' interests and biases.

Second, develop a taxonomy of uncertainty. It helps to establish a framework for modellers and rule makers to understand and analyse the kinds of uncertainty with which they are coping. This taxonomy clarifies different kinds of risks and uncertainties associated with mathematical modelling and embedded in crises. Although mathematical models can minimize the complexity of the real world and give precise answers in numbers, their role is limited under conditions of uncertainty, as not all kinds of uncertainty can be described as well-defined, numerical probabilities. If rule makers approach mathematical modelling as 'truth' and manipulate it as 'evidence' to support predetermined agendas under conditions of radical uncertainty, their approach is flawed, and the public cannot really know what works and how to effectively address the challenge. Reliance on mathematical models may also downplay other sources of knowledge and expertise.

For example, experts argued that in coping with the COVID-19 pandemic, the UK government gave too much weight to the views of modellers, while overlooking the views of public health experts (Devlin and Boseley 2020).

Finally, re-evaluate the nature and function of ignorance and its importance for examining the interaction between mathematical modelling and rule-making. This supports the view that ignorance is a condition of knowledge and argues that rather than eliminating uncertainty or hiding ignorance behind their expertise, modellers and rule makers should embrace uncertainty and ignorance. They should mobilize ignorance in a productive way through finding solutions to problems that they failed to identify earlier.

The new strategy to number- and model-based rule-making includes three essential aspects, corresponding to the three steps set out in this chapter. First, modellers and rule makers need to reflect on the nature of mathematical modelling, including how data is gathered and how information is captured. They should ask: what data is missing? What factors are not considered? What assumptions are made in the process of mathematical modelling? Second, modellers and rule makers should assess the kind of uncertainty with which they are coping. Third, modellers and rule makers need to communicate the unknown in decision-making. In communicating, they should not conceal fear, doubt, or dissent behind claims of truth and absolute control made through the outputs of mathematical modelling. Instead, they should embrace and work with uncertainty, acknowledging that our ignorance is infinite. Recognizing the virtue of their awareness of ignorance encourages modellers and rule makers to find missing data and listen to unheard voices that they and other experts failed to identify earlier. Learned ignorance pushes them to find solutions to crises but also to accept that not all is knowable.

References

Adam, D. 2020a. 'Special Report: The Simulations Driving the World's Response to COVID-19'. *Nature*, 580, 316–18.

Adam, D. 2020b. 'Simulating the Pandemic: What COVID Forecasters Can Learn from Climate Models'. *Nature*, 587, 533–4.

Beck, U. 1992. *Risk Society: Towards a New Modernity*, translated by M. Ritter, Sage Publications, London.

Best, J. 2010. 'The Limits of Financial Risk Management: Or What We Didn't Learn from the Asian Crisis'. *New Political Economy*, 15(1), 29–49.

Chivers, T. 2021. 'The Obscure Maths Theorem That Governs the Reliability of Covid Testing'. *The Observer*, 18 April 2021. https://www.theguardian.com/world/2021/apr/18/obscure-maths-bayes-theorem-reliability-covid-lateral-flow-tests-probability?CMP=Share_AndroidApp_Other (last accessed 1 April 2023).

Davies, W. and L. McGoey. 2012. 'Rationalities of Ignorance: On Financial Crisis and the Ambivalence of Neo-liberal Epistemology'. *Economy and Society*, 41(1), 64–83.

Devlin, H. and S. Boseley. 2020. 'Scientists Criticise UK Government's "Following the Science" Claim'. *The Guardian*, 23 April 2020. https://www.theguardian.com/world/2020/apr/23/scientists-criticise-uk-government-over-following-the-science (last accessed 1 April 2023).

Drechsler, W. 2011. 'Understanding the Problems of Mathematical Economics: A "Continental" Perspective'. *Real-World Economics Review*, 56, 45–57.

Ferguson, N.M. et al. 2020. 'Report 9: Impact of Non-Pharmaceutical Interventions (NPIs) to Reduce COVID-19 Mortality and Healthcare Demand'. 16 March 2020. https://www.imperial.ac.uk/mrc-global-infectious-disease-analysis/covid-19/report-9-impact-of-npis-on-covid-19/ (last accessed 1 April 2023).

Friedman, J. 2005. 'Popper, Weber, and Hayek: The Epistemology and Politics of Ignorance'. *Critical Review*, 17(1–2), 1–58.

Funtowicz, S. and J.R. Ravetz. 1993. 'Science for the Post-Normal Age'. *Futures*, 31(7), 735–55.

Funtowicz, S. and A. Saltelli. 2015. 'Evidence-Based Policy at the End of the Cartesian Dream: The Case of Mathematical Modelling'. In *Science, Philosophy and Sustainability: The End of the Cartesian Dream*, edited by Â. Guimarães Pereira and S Funtowicz, Routledge, Abingdon, pp. 147–62.

Giddens, A. 1999. 'Risk and Responsibility'. *Modern Law Review*, 62(1), 1–10.

Giddens, A. and C. Pierson. 1998. *Making Sense of Modernity: Conversations with Anthony Giddens*, Stanford University Press, Stanford, CA.

Harford, T. 2020. 'Why It's too Tempting to Believe the Oxford Study on Coronavirus'. *Financial Times*, 27 March 2020. https://www.ft.com/content/14df8908-6f47-11ea-9bca-bf503995cd6f?desktop=true&segmentId=7c8f09b9-9b61-4fbb-9430-9208a9e233c8#myft:notification:daily-email:content (last accessed 1 April 2023).

Hayek, F.A. 1945. 'The Use of Knowledge in Society'. *The American Economic Review*, 35(4), 519–30.

Holmdahl, I. and C. Buckee. 2020. 'Wrong but Useful: What Covid-19 Epidemiologic Models Can and Cannot Tell Us'. *The New England Journal of Medicine*, 383, 303–5.

Horton, R. 2020. 'Offline: COVID-19: A Reckoning'. *The Lancet*, 395, 935.

Jewell, N.P., J.A. Lewnard, and B.L. Jewell. 2020. 'Predictive Mathematical Models of the COVID-19 Pandemic: Underlying Principles and Value of Projections'. *JAMA*, 19, 1893–4.

Johns Hopkins Center for Health Security. 2019. 'Global Health Security Index: Building Collective Action and Accountability'. https://www.ghsindex.org/wp-content/uploads/2019/10/2019-Global-Health-Security-Index.pdf (last accessed 1 April 2023).

Johns, F. 2020. 'Model the Pandemic, Model the World: Which World? Whose World?' *ILA Reporter (The Official Blog of the International Law Association (Australian Branch))*, 22 December 2020. https://ilareporter.org.au/2020/12/model-the-pandemic-model-the-world-which-world-whose-world-fleur-johns/ (last accessed 1 April 2023).

Kay, J. and M. King. 2020. *Radical Uncertainty: Decision-Making Beyond the Numbers*, The Bridge Street Press, London.

Keynes, J.M. 1937. 'The General Theory of Employment'. *Quarterly Journal of Economics*, 51(2), 209–23.

Knight, F. 1921. *Risk, Uncertainty and Profit*, Courier Corporation, Boston and New York.

Lawson, D. and G. Marion. 2008. 'An Introduction to Mathematical Modelling'. https://people.maths.bris.ac.uk/~madjl/course_text.pdf.

McCloskey, D. 2002. *The Secret Sins of Economics*, Prickly Paradigm Press, Chicago, IL.

McGoey, Linsey. 2012. 'Strategic Unknowns: Towards a Sociology of Ignorance'. *Economy and Society, 41(1)*, 1–16.

Merry, S.E. 2016. *The Seductions of Quantification: Measuring Human Rights, Gender Violence, and Sex Trafficking*, University of Chicago Press, Chicago, IL, and London.

Murphy, T. 1997. *The Oldest Social Science? Configurations of Law and Modernity*, Oxford University Press, Oxford.

Nelken, D. and M. Siems, eds. 2021. 'Numbers in an Emergency: The Many Roles of Indicators in the COVID-19 Crisis'. *International Journal of Law in Context, 17(2)*, 161–7.

Nietzsche, F. 2003 (1973). *Beyond Good and Evil: Prelude to a Philosopher of the Future*, translated by R.J. Hollingdale, Penguin Books, London.

Perry-Kessaris, A. 2011a. 'Prepare Your Indicators: Economics Imperialism on the Shores of Law and Development'. *International Journal of Law in Context, 7*, 401–21.

Perry-Kessaris, A. 2011b. 'Reading the Story of Law and Embeddedness Through a Community Lens: A Polanyi-Meets-Cotterrell Economic Sociology of Law?' *Northern Ireland Legal Quarterly, 62(4)*, 401–13.

Perry-Kessaris, A. 2017. 'The Re-co-construction of Legitimacy of/through the Doing Business Indicators'. *International Journal of Law in Context, 13*, 498–511.

Popper, K. 1985. *Popper Selections*, translated by D.W. Miller, Princeton University Press, Princeton, NJ.

Ravetz, J.R. 1987. 'Usable Knowledge, Usable Ignorance'. *Knowledge, 9(1)*, 87–116.

Ravetz, J.R. 2015. 'Preface: Descartes and the Rediscovery of Ignorance'. In *Science, Philosophy and Sustainability: The End of the Cartesian Dream*, edited by Â. Guimarães Pereira and S. Funtowicz, Routledge, Abingdon, Oxon, pp. xv–xviii.

Rayner, S. 2012. 'Uncomfortable Knowledge: The Social Construction of Ignorance in Science and Environmental Policy Discourses'. *Economy and Society, 41(1)*, 107–25.

Reinert, E.S. 2009. 'The Terrible Simplifiers: Common Origins of Financial Crises and Persistent Poverty in Economic Theory and the New "1848 Moment"', DESA Working Paper No. 88, https://www.un.org/esa/desa/papers/2009/wp88_2009.pdf (last accessed 1 April 2023).

Reinert, E.S., M. di Fiore, A. Saltelli, and J. Ravetz. 2021. 'Altered States: Cartesian and Ricardian Dreams', UCL Institute for Innovation and Public Purpose, Working Paper Series (IIPP WP 2021/07), http://www.ucl.ac.uk/bartlett/public-purpose/wp2021–07 (last accessed).

Rhodes, T. and K. Lancaster. 2020. 'Mathematical Models as Public Troubles in COVID-19 Infection Control: Following the Numbers'. *Health Sociology Review, 29(2)*, 177–84.

Rodrik, D. 2015. *Economic Rules: Why Economics Works, When It Fails, and How to Tell the Difference*, Oxford University Press, Oxford.

Rottenburg, R., S.E. Merry and S.-J. Park, eds. 2015. *The World of Indicators: The Making of Governmental Knowledge through Quantification*, Cambridge University Press, Cambridge.

Saltelli, A., G. Bammer, I. Bruno, E. Charters, M. Di Fiore, E. Didier, W. Nelson Espeland, J. Kay, S. Lo Piano, D. Mayo, R. Pielke Jr, T.M. Portaluri, T. Porter, A. Puy, I. Rafols, J.R. Ravetz, E. Reinert, D. Sarewitz, P.B. Stark, A. Stirling, J. van der Sluijs, and P. Vineis. 2020. 'Five Ways to Ensure that Models Serve Society: A Manifesto'. *Nature, 582*, 482–4.

Scoones, I. and A. Stirling. 2020. *The Politics of Uncertainty: Challenges of Transformation*, Routledge, London.

Smithson, M. 1989. *Ignorance and Uncertainty: Emerging Paradigms*, Springer-Verlag, New York.

Snow, T. 2021. 'The (Il)logic of Legibility: Why Governments Should Stop Simplifying Complex Systems', *LSE Impact Blog*. 12 February 2021. https://blogs.lse.ac.uk/impactofsocialsciences/2021/02/12/the-illogic-of-legibility-why-governments-should-stop-simplifying-complex-systems (last accessed 1 April 2023).

Spiegelhalter, D. 2019. *The Art of Statistics: Learning from Data*, Pelican, London.

Stewart, T.R. 2000. 'Uncertainty, Judgment, and Error in Prediction'. In Prediction: Science, Decision Making, and the Future of Nature, edited by D. Sarewitz, R.A. Pielke, and R. Byerly, Island Press, Washington, DC, pp. 41–60.

Supiot, A. 2017. *Governance by Numbers, The Making of a Legal Model of Allegiance*, translated by S. Brown, Hart Publishing, Oxford.

Taleb, N.N. 2010. *The Black Swan: The Impact of the Highly Improbable*, 2nd edition, Random House, London.

Whitty, C.J.M. 2015. 'What Makes an Academic Paper Useful for Health Policy?' *BMC Medicine, 13(301)*, 1–5.

Winner, L. 1980. 'Do Artifacts Have Politics'. *Daedalus, 109(1)*, 121–36.

11

In the twilight of probability

COVID-19 and the dilemma of the decision maker

Paolo Vineis and Luca Savarino

Introduction

'In the twilight of probability' is an expression used by John Locke in his essay 'Two Treatises of Government' to remind us that certainty is not a common experience in this world:

> Therefore, as God has set some things in broad daylight; as he has given us some certain knowledge, though limited to a few things in comparison, probably as a taste of what intellectual creatures are capable of to excite in us a desire and endeavour after a better state: so, in the greatest part of our concernments, he has afforded us only the twilight, as I may so say, of probability; suitable, I presume, to that state of mediocrity and probationership he has been pleased to place us in here.
>
> (Locke 1689)

Probabilistic knowledge is the norm in science and in everyday life; certainties are few. There is also another way to acknowledge our constant state of uncertainty, which is more strictly epistemological: the myth of the cave by Plato. Like the prisoners in the myth, we only see shadows on a wall—that is, we do not have direct access to things 'as they are'. Uncertainty permeates the interpretation of the past and—arguably more so—anticipation of the future, or prediction. In this chapter, we will consider the role of mathematical modelling in both the context of scientific uncertainty and the broader context of value-laden decision-making.

The Anthropocene, uncertainty, and prediction

Nowadays, things are much more complicated than in Plato's or Locke's time. In the early 2000s, Paul Crutzen and Eugene Stoermer coined the term 'Anthropocene'. The word refers to the era in which human beings have become able to

Paolo Vineis and Luca Savarino, *In the twilight of probability*. In: *The politics of modelling*. Edited by: Andrea Saltelli and Monica Di Fiore, Oxford University Press. © Oxford University Press (2023). DOI: 10.1093/oso/9780198872412.003.0011

transform the entire habitat of the planet so radically as to affect even deep geological strata (Crutzen and Stoermer 2000). There is a lack of agreement on the exact date when the Anthropocene actually started, the most widely accepted view being that it goes back to the invention of the steam engine and the industrial revolution. Whatever the date of origin, in the twenty-first century the term quickly became popular to explain the environmental crisis. The pandemic can also be read within this framework. Indeed, there is a deep link between climate change and the progressive spread of infectious diseases in the past few years. Factors such as global warming, deforestation, loss of biodiversity, and the exponential increase in intensive animal farming combined to create the conditions conducive to the spread of pathogens and an increased risk of zoonotic events.

In early 2020, we kept asking ourselves nearly obsessively whether the pandemic was the product of *fate* or *fault*: in other words, whether it was an unforeseeable natural catastrophe or the result of human choices that *could* have been avoided or *should* have been foreseen. A clear-cut answer is far from being evident, because the current features of the planet make the use of the tools of modern Western philosophy, based on the distinction between nature and culture, almost impossible (Descola 2013). Aurelio Peccei's words from 50 years ago sound prophetic: we live in 'a small world, increasingly dominated by interdependencies that make it an integrated global system where humans, society, technology and Nature condition each other through increasingly binding relationships' (Meadows et al. 1972)[1]. The Anthropocene reality can only be described as an increasingly *cultured nature*, one constantly reshaped and destabilized by the action of human beings. At the same time, it can also be described through the image of a global culture where human life is governed by socioeconomic mechanisms that appear to be as inescapable as *nature*'s laws. The same is true for the SARS-CoV-2 virus which, according to the nature–culture mental frame of reference, is only too often referred to *either* as a natural event—that has come to us from the depths of the Asian forests—*or* as an artefact of the Wuhan laboratories. In fact, COVID-19 is not well suited to either of these definitions: it resembles a network, or a *hybrid*, to refer to Bruno Latour's definition (Latour 1993).

One theoretical framework that may be able to deal with the study of pandemic-related issues is that of a *hyperobject*, suggested by the British philosopher Timothy Morton (2013). Although it is not easy to define what a hyperobject is in theory, its meaning is intuitively clear: it is a boundless, outsized object; you can perceive its local effects, but it escapes objectification, because it is something we are *in* rather than being *opposite to* or *in front of* it. One of its specific characteristics is that it is spatially *non-local* and *time phasic*. For instance, the pandemic, demographic pressure, and climate change generate very different results according to

---

[1] The quote is taken from the Preface to the Italian edition of Meadows et al. 1972 (*I limiti dello sviluppo*, Edizioni Scientifiche e Tecniche Mondadori, Milano, pp. 13–14). Our translation.

the different areas of the planet they affect, and it is very hard to predict the shape of the geographic distribution of their effects in the future. The spread of the SARS-CoV-2 virus, and the way viruses interact with host organisms, how they mutate and circulate, etc., implies a timeline of the evolution of the pandemic with temporary regressions and sudden accelerations (that is, *phases*), making an accurate quantification near impossible. Precisely because of the reticular or networked nature of the epidemic—that is, the lack of a clearly defined centre from which to track and measure causal chains—it is very difficult to act effectively according to a predefined plan.

It has been repeated several times that the COVID-19 pandemic could have been foreseen, and in fact something similar ('disease X') had been predicted (McKenzie 2020). However, when we become indignant about the lack of preparedness, we should not forget that such predictions are general, if not vague: where will the next pandemic start? How will it be spread? How lethal? Nobody can tell with certainty.

If we discard the nature–culture dichotomy and adopt the point of view proposed above, the answer to the '*fate* or *fault*' question becomes more nuanced—that is, the responsibilities of the pandemic are partly social and partly cultural. COVID-19 is the result of a series of human actions on the planet, partially overlapping with the same actions that caused the current environmental crisis. However, while knowing that something like this was likely to happen, we did not know *when* it was going to take place. In fact, the network of causes has become increasingly complicated: i) natural and social causes mix together and do not allow an easy identification of root causes or primary events, and ii) causal chains expand beyond limits, highlighting distal causes we cannot usually observe. Furthermore, a degree of fortuitousness in the field increases the chances of pathological events occurring, which is true not only for infectious but also for chronic diseases. Diseases such as diabetes or cancer are not just single-cause illnesses, but are the result of more complex processes, due to a combination of exposures, none of which is necessary or sufficient in itself. For a long time, we were used to thinking of disease as something to be addressed only when symptoms became obvious, and as having a proximal and perhaps necessary cause (think of the classical infectious diseases). The progress of science has made us aware of a much greater level of complexity: causes are distal and belong to chains of causation; diseases also exist in the form of asymptomatic conditions; and what we considered fate was a mixture of this causal complexity and of human and social responsibility.

Coming back to the pandemic, even the tools created to monitor it, including prediction models, as we shall see, reflect the pandemic as a *hyperobject*. Consider R_0, which is obviously the result of our conceptualization via models—that is, an attempt to capture the *hyperobject* by simplifying it. As argued by some (Miller 2022), R_0 is not an observable entity in itself, but the product of a model: since the pandemic, more than before, we have interiorized model thinking, a trend also reflected in many other manifestations of current societies. R_0, the 'repro-duction number' that characterizes the spread of a virus in a population, varies from <1 for Middle East Respiratory Syndrome (MERS) or 1–2 for seasonal flu, up to 12–18 (that is, one contagious child infects 12–18 others) for measles. And lethality varies from 25–90% for Ebola down to 0.1% for flue and measles. On this basis, it is impossible to predict which characteristics the next epidemics will have: it will depend on the types of mutations viruses have undergone (which in turn will depend on the role of original and intermediate animal reservoirs), the degree of susceptibility of the population and their admixture (the matrix of contacts), the modalities of transmission, etc. The ensuing *humility* does not imply inactiv-ity. If it is true that the frequency of pandemics will increase, as several scientists claim (Morens and Fauci 2020, Carlston et al. 2022), we need to address the roots of such an increase, while attempting to block epidemics with surveillance sys-tems. And in fact, fortunately, not everything occurs in the twilight of probability: we also have certainties. Given the difficulties of prospective predictions, it is even more exciting that we can count on retrospective spectacular successes: in 1967, 2,000,000 deaths were due to smallpox and 2,500,000 to measles, while today they are 0 and less than 100,000 per year respectively, thanks to vaccines. It has been estimated that the measles vaccine has avoided 20.3 million deaths in the period 2000–15 (Patel 2016). Only a small proportion of children in the world are vac-cinated against measles, which explains the partial success. We have certainties, after all.

We could say that knowledge has a continuous distribution from 0 (for phenom-ena we know nothing about, like unknown viruses to come) to 1 (for events like the existence of Nazi concentration camps or the laws of macrophysics), with most judgements floating in the twilight of probability. This is at the root of a decision maker's dilemma: scientific evidence is distributed along this continuum, most of it being provisional and uncertain.

Mathematical modelling

The great medical historian Mirko Grmek coined the term 'pathocenosis' to describe the relationship between human diseases and the surrounding environ-ment in its historical determinations (Grmek 1969). These relationships are largely based on the role of the immune system and the combination of mutation and selection of viral variants. The fact that an infectious disease can cause a pan-demic depends on two main conditions: the state of immunological susceptibility to infection of the entire population (a condition that occurs, for example, when a virus is new to humans) and the aetiological agent's ability to transmit itself effi-ciently from one person to another (and this is the more circumstantial part). The emergence of a new virus or bacterium is not a rare occurrence, but is part of normal evolutionary processes. The ability to infect new animal species, including humans, gives these mutated viruses or bacteria a selective advantage, as they are able to expand into new ecological niches. The impact of a pandemic on the population depends on the spread rate, the severity of the clinical picture, and the lethality rate. In mathematical modelling, the spread rate is measured by a basic reproduction index (R_0)—that is, the average number of people infected (secondary cases) in a population that is fully susceptible to every single contagious case. The index is directly proportional to the duration of contagiousness of the infected person and the frequency of contacts during which other people are exposed to the infection. It is therefore understandable that social distancing measures are effective by reducing one of the parameters of R_0. Several infectious agents, even common ones, have a very high reproduction index under conditions of complete susceptibility of the population. For example, for measles it is usually estimated that an average of 12–18 secondary cases occur for each primary case (Guerra et al. 2017). However, for most circulating infections, the majority of the population has now developed immunity, either because people have already contracted the infection or because they have been vaccinated against it. The proportion of people immune to a specific infection in the population is a hindrance to the spread of the infection, as every immune person—even if exposed to infection—supposedly will not infect anyone else. When the proportion of immune persons in a population is so high that it does not allow for epidemic spread, it is said that the population has developed 'herd immunity'.[2] Most of the determinants and characteristics of the spread of the infection in a population seem to be *circumstantial, non-deterministic,* and *local.*

The severity of the clinical picture is another important parameter regarding the relevance of the pandemic and its spread. Infections characterized by a severe clinical picture are more easily detected and, if the contagiousness is limited to the period of time in which the symptoms occur, they are easier to intercept. These aspects may explain why some infections cause pandemics and others only epidemics, which remain circumscribed. The SARS epidemic of 2003, for example, was more contained than the current spread of COVID-19: the R_0 was similar, but the mortality rate was significantly higher (9–16%), and above all, the highest infection rate was found in the second week after the onset of symptoms. These characteristics facilitated the identification of cases and the isolation of people exposed to the infection (their contacts), leading to the eradication of the epidemic before it spread broadly like COVID-19.

Indeed, every epidemic has its own characteristics, which are linked to the type of aetiological agent, the harm it induces, and the way it is transmitted, so that it is difficult to derive counter-measures from one epidemic to another and predict

---

[2] A questionable term: humans are not a herd—that is, they have a much more complex social structure and autonomy than animals.

their course by analogy. The behaviour of different pathogens can be estimated by constructing mathematical models that simulate the conditions of infection transmission, produce spread scenarios, and offer the possibility of evaluating the effect of specific counter-measures. Therefore, the history of the circulation of each epidemic or pandemic has a scientific reproducible component, in the sense that it can be interpreted in the light of the virus's mutations and adaptation to the host, and the latter's immune response. But the overall narrative has many circumstantial and unpredictable elements, linked to chance (for example, the appearance of the right mutation at the right time), and the geographical and historical context. The term *pathocenosis* encompasses these aspects as well as the unstable balance between different diseases in a population; the pathocenosis is constantly in a condition of precarious balance or imbalance. It should be noted that the human (or animal) response to microbial aggression also leads to a constant instability of the immune system, which is subject to continuous recombinations of its genetic material for the production of antibodies corresponding to new environmental antigens.

This complexity has characterized the pandemic, partly as an effect of the Anthropocene, and has led to a different perception of causality and its relationships to responsibility. If we go back to the roots of the concept of cause, it was originally intertwined with that of 'guilt': in ancient Greek they were the same word (*aitia*), which led to a long tradition of 'search for the culprit' of events. Even recently, single culprits have been looked for in the case of COVID-19 (for example, the Chinese, rather than the global crisis). This is typical of current populist politics (Trumpism), and of conspiracy theories. However, most of us have eventually understood that single causes operate within networks, or webs of causation, and this is certainly a great achievement of modern science and goes together with the 'twilight of probability'. This evolution also has a moral downside—that is, we cannot always identify a single responsible agent (though in many cases we can), but events are the product of complex networks and therefore derive from collective responsibilities. Again, we have a continuum from 0 (no one is responsible, as in the case of earthquakes) to 1 (a murderer), with an intermediate set of grey situations that represent a mixture of individual and collective responsibilities, which is the most common situation.

When it comes to mathematical models, humility has been invoked (Jasanoff 2007), in the sense that models are affected by uncertainties and cannot be stretched beyond their limits of prediction. Let us consider just a couple of examples.

The first refers to the early predictions made by Neil Ferguson's team at Imperial College, which prompted the adoption of non-pharmacological containment measures in several countries.[3] The issue is that the models were inevitably based

---

[3] See 'Report 9: Impact of Non-Pharmaceutical Interventions (NPIs) to Reduce COVID-19 Mortality and Healthcare Demand', https://spiral.imperial.ac.uk/bitstream/10044/1/77482/14/2020-03-16-COVID19-Report-9.pdf (last accessed 2 April 2023).

on limited experience (first data from Wuhan, then from Northern Italy), plus a number of assumptions: (a) they assumed an incubation period with mean 5.1 days and standard deviation 4.4 days, estimated from travellers' case data (this turned out to be correct); (b) infectiousness was assumed to start 0.5 days before symptom onset (also correct, with individual variability); (c) they assumed that individuals vary in infectiousness according to a gamma distribution with mean 1 and dispersion parameter k=0.25 (though other, later estimates suggested k=0.1; this means that 10% of the cases give origin to 80% of infections). This level of overdispersion matched estimates obtained from observed transmission chains; in other words, they assumed that there were overspreaders of disease, which also turned out to be correct; (d) they assumed R_0 = 2.4, as estimated in Wuhan. In fact this was a high value, which later became lower and was substituted by Rt—that is, the reproduction number influenced by the containment measures that were taken; (e) they assumed that 50% of all infections were symptomatic[4] (at least mildly). The lesson from this example is that early modelling is necessarily tentative, based on fragile assumptions, and requires constant updates.

A second example is a concrete application of modelling for policy-making in Piedmont, Italy. Figs. 11.1 and 11.2 show a 'SEIRS' model—that is, a mathematical model that assumes the existence of compartments in the population (Susceptible-Exposed-Infected-Removed-Susceptible) and transition rates between them. The model uses assumptions on the matrix of contacts—that is, how frequent contacts are in the population, depending on the stage of the epidemic and containment measures taken (Fig. 11.1)—and makes predictions (Fig. 11.2) on the development of the epidemic curve that depend entirely on the assumptions made on the effectiveness of devices and compliance with social distancing.

These are some examples of why we have concluded in *Manifesto* (Saltelli et al. 2020) that modellers should not project more certainty than their models deserve; and politicians must not be allowed to offload accountability to models of their choosing.

The uncertainties in data collection and mathematical modelling for COVID-19 were expressed in early phases of the epidemic, for example by Waltner-Toews et al. (2020):

> Known unknowns [in the present pandemic] include: the real prevalence of the virus in the population; the role of asymptomatic cases in the rapid spread of the virus; the degree to which humans develop immunity; the dominant exposure pathways; the disease seasonal behaviour; the time to global availability of an effective vaccine or cure; and the nonlinear response of individuals and collectives to the social distancing interventions in the complex system of communities

---

[4] This is still highly uncertain, depending on the design of the investigations; therefore, estimates are very variable across populations and infection waves.

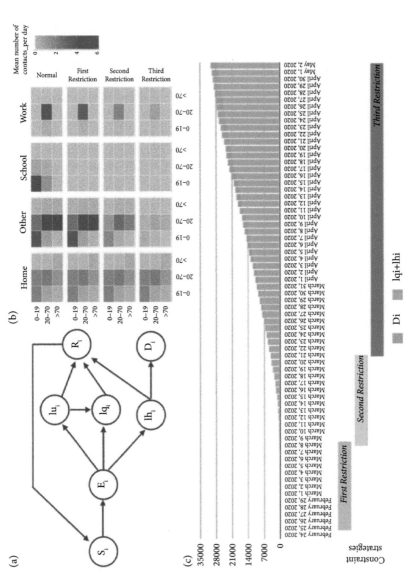

Fig. 11.1 SEIRS (Susceptible-Exposed-Infected-Removed-Susceptible) model and surveillance data in the Piedmont region: a) The transmission flow diagram of our age-dependent SEIRS model. b) Age-specific and location-specific contact matrices. The intensity of the colour indicates a higher propensity to make contact. c) Distribution of infected cases as sum of quarantined (Iqi), hospitalized (Ihi), infected (light green), and deaths (Di) (dark green) from 24 February to 2 May 2020. The periods of the activation of three control strategies are reported below the stacked bars plot.

Source: Pernice et al. 2020. *BMC Infectious Diseases*, 20(1), 798.

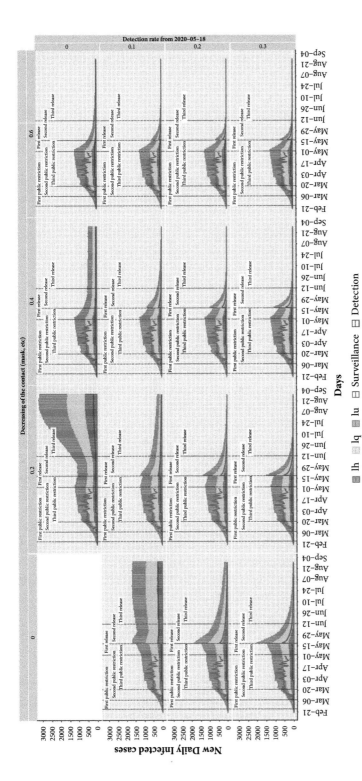

Fig. 11.2 Combinations of the different levels of effectiveness of individual protection devices and distancing rules (0%, 20%, 40%, and 60%) (columns) and the different levels of capacity of identifying infected individuals (0%, 10%, 20%, and 30%) (rows).

Source: Pernice et al. 2020, *BMC Infectious Diseases,* 20(1),798.

interconnected across multiple scales, with many tipping points, and hysteresis loops (implying that society cannot re-bounce to its pre-corona interventions state).

Mathematical models produce uncertain numbers that predict future infections, hospitalizations, and deaths under various scenarios. Only some of these uncertainties have been reasonably resolved in later phases, but in any phase of the epidemic development, numbers can be used politically in an inappropriate way (Rhodes and Lancaster 2020). For example, rather than using models to inform their understanding, political rivals often brandish them to support predetermined agendas.

These issues are only partially addressed by model validation and verification approaches. Some of the issues are purely technical—such as, for example, in using sensitivity analysis, or sharing the code in repositories—while others are those which we called reflexive (Saltelli et al. 2020), and look at a richer set of attributes of the modelling process, including contextual framings, expectations, and motivations of the developers, tacit assumptions, implicit biases in the use of a particular modelling approach, care in not overinterpreting the results, and so on. An example of a technical prescription for model quality in the case of COVID-19 is sharing a well-documented and interpretable version of the algorithm code of predictive models. While this is certainly a useful recommendation, its scope does not match the vast scale of the problem of prediction.

Mathematical models are both descriptive and performative—that is, they influence society and are influenced by this in their becoming, as shown by many examples ranging from economics to ecology, as well in the pandemic. Discussing COVID-19 models, Rhodes and Lancaster (2020) speak of 'mathematical models as public troubles', but conclude on a positive note that these troubles 'can be treated as generative events for energising new forms of expertise, as well as for remaking experts and publics, through participation'. Sometimes precision of estimates serves the purpose of persuading politicians to act, but it is also a double-edged sword. For example, in the case of the Imperial College model, which played an important role in the adoption of policies both in the UK and US, the predictions of 510,000 COVID-19-related deaths in the UK and 2.2 million in the US were too precise given the uncertainty in model inputs, as was the authors' own limited assessment of the uncertainties (a range between 410,000 and 550,000), achieved by looking at the dependence of the results on the uncertain value of infection rate.[5]

An attitude inspired by *docta ignorantia* has been suggested by, for example, Jerome Ravetz: an appraisal of ignorance can allow the detection of 'trans-scientific' problems—that is, problems that can be formulated in the language of

---

[5] Additional examples relevant to COVID-19 and other epidemics are described in the Supplementary Material of Saltelli et al. 2020.

science but to which science cannot offer a solution. In such cases, quantification is either superfluous or counterproductive, as expressed by Anthony Fauci's famous statement: 'There is no number-answer to your question.'

Facts and values: On science and politics

The problem we have discussed so far, which concerns uncertainties in the development and use of mathematical models, is not the only one. In fact, policy-making also refers to values, not only to facts and predictions. This shifts the argument substantially: even if we can imagine having perfectly predictive models with little margin for uncertainty, still the problem of finding agreement on values in policy-making would persist. Typically, a decision maker has to find a balance between evidence and values, though the temptation of politicians is to delegate decisions to scientists, which is the easiest way to avoid delving into the traps of value systems. In the first wave of COVID-19, at least in Italy, politicians transferred into action the scientists' point of view and guidelines, something unusual for them. The decision pathways became much more tortuous after the summer of 2020, when it became clear that either science was not taken seriously (many politicians, but also some clinicians preferred to believe that the epidemic was over, against all likelihood) or politicians had to incorporate value-laden aspects into their decisions. For example, the UK led the way in delivering guidelines for vaccination based on frailty, which imposed a hierarchical sequence based on age and clinical susceptibility. This was a value-laden decision that gave priority to the elderly over the young, and was dictated by health rather than economic reasons. In Italy, the process was the opposite at first—that is, many regional presidents started vaccination campaigns that lacked coherence, to say the least, and in fact gave priority to economic reasoning, particularly of certain economic sectors.

In practice, it is impossible to rely exclusively on science, and all policy decisions are inevitably a mixture of evidence and values. Science helps us to calculate the consequences of our actions, but it is in itself insufficient, because the political question is not really about facts or even values, but rather about the relationship between facts and values. We can say that *politics or policy-making is the art of finding a balance between different values (and interests) on the basis of the scientific evidence available at a given time* (see also Gluckman 2014). This definition condenses all the difficulties and uncertainties that have been experienced at a political level during the management of the pandemic.

Putting it in this way, we can perhaps interpret in a different light some of the events that characterized the pandemic. We can, for example, unearth hidden values in certain public statements. Although it was numerically almost irrelevant, the point of view of representatives of the galaxy against vaccines and the 'green pass' (used to check the immunity status of people) in Italy was very vocal and amplified by the media. Some leaders of the movement tried to deny that the scientific evidence justified the measures that were taken, and claimed that the vaccine was ineffective or even that it caused serious side effects. Behind the attempt to use (faulty) scientific arguments, part of the galaxy can be defined as 'ultraliberal', reminiscent of radical libertarian thinkers like Robert Nozick, who championed the shrinking of the state to a minimum, or Thatcher's famous statement 'There's no such thing as society.'

In fact, many policy decisions made during the pandemic were implicitly value-laden, though the rationale was not discussed in moral terms, but there was an attempt to justify them through scientific 'necessity' or claiming that the basis was self-evident. The most obvious of such decisions 'in the twilight of probability' was closing schools down. There is uncertain evidence that closing schools was useful for fighting the pandemic, while there is ample evidence that remote teaching was clearly harmful. By the way, one of the great difficulties in taking such decisions is that we should apply something equivalent to the discount rate for investments—that is, take into account the long-term effects. The lockdown of all activities (except essential ones) has an immediate effect on mortality among the elderly and the most susceptible, and is correct on the basis of a Kantian value system—one which treats all equally (except supermarket cashiers or bus drivers). But—in utilitarian terms—the Kantian approach does not count the long-term effects—including on health—related to the impact on education and the economic system. Someone will sooner or later provide such (difficult) calculations, but they are clearly unfeasible when a rapid decision needs to be taken. In practical terms, we do not suggest that policy makers should specify—at the time of tragic choices—whether they are libertarian, Kantian, or utilitarian, but rather involve the society in a long-term debate 'in peace times'—that is in the interval between epidemics—on how conflicting values should be dealt with and included in political decisions (Delanty 2020).

In history, difficult relationships between science and politics abound, with many political intrusions into science-driven decisions that have clear similarities with the present situation. For example, in Hamburg in 1892, with six weeks 8600 people died from cholera because authorities did not want to take containment measures so as to avoid damaging commercial activities in the harbour. This was related to a strong autonomist political tendency embedded in the Hanseatic League, which was hostile to the Prussian bureaucracy. It was Robert Koch who eventually imposed containment measures on behalf of the emperor. Epidemics have also played a significant role in geopolitics in the past. The opening of the Suez Canal (1869) was followed by measures by an international commission to control the spread of cholera from India. However, quarantines of ships were interpreted by the British as a violation of the Free Trade Act perpetrated by the French and a manifestation of trade wars.

Inseparable distinction?

The relationships between science and policy-making have been extensively explored in philosophy and social science (see, for example, Majone 1989). One famous and still relevant example is Max Weber with his lectures on science and politics as vocations (Weber 2004). Weber stresses two simultaneous attitudes of modernity towards truth. On the one hand, science teaches us not to search for a 'thing as such' ('Ding an Sich'), and encourages us to embrace the 'disenchantment of the world'. On the other hand, modern liberal politics liberate us from an idea of authority that embodies some kind of moral truth. The term 'disenchantment of the world' refers to that process of rationalization of society, resulting from the development of technology and economic capitalism, which substitutes conventional relations for the organic and communal ties typical of traditional societies. In particular, the birth of modern science marks the establishment of a conception of nature as a unitary mechanism, governed by cause-and-effect relationships, which functions according to laws that can be used to calculate and thus to predict.

One consequence of the emergence of the modern scientific mentality is precisely the dissociation between facts and values. Both on the side of evidence and on the side of values, we are left with a disenchantment which is at the root of the constant uncertainty. Science explains the consequences of a specific choice, but is unable to justify whether that choice is beneficial to human ends and needs. Medicine in itself does not tell us whether prolonging life or limiting suffering are values to pursue, particularly when they are in conflict, though it can describe in detail the mechanisms that lead one way or the other. The realm of the ends, as opposed to the means, is outside the goals of modern science, as Weber stresses. On the other side, a completely rational approach to values—similar to the scientific one—is impossible. When public choices are involved, Weber suggests that we are inevitably surrounded by a 'polytheism' of (conflicting) values, and this cannot be solved by the authority principle in a liberal society. The scientist does not have to express ideas about ends, but must limit himself to describing and comparing worldviews and indicating their assessment of the consequences that may follow, with associated uncertainties, without establishing their value, which remains by definition outside the scope of the investigation of science. The ethics of responsibility as the ability to predict rationally the consequences of one's own actions remains distinct from the ethics of principles, which refers to values. In the modern world, the choice about the end is entrusted, even before individual freedom, to the political contention that determines which of the many available values should be pursued by a given society.

Weber speaks of 'polytheism of values' to describe the mutual struggle between deities who express irreconcilable views on the ultimate meaning of life: what science is unable to offer is a rational decision about which 'god' to serve; ends and values are not universal, because they are not justifiable rationally. The difference between the traditional conception and the modern conception of science is precisely this: while the former believes that science can grasp the ultimate meaning of the world—understood at different times as 'true being, true nature, true God, true happiness'—modern science is constitutively incapable of tapping into the realm of ends.

We are convinced that the importance of Weber's teaching consists precisely in the idea of the impossibility of reducing politics to technology. However, just as we are facing radical changes in the planet (the Anthropocene, climate change, the food crisis), the concept of Weber's 'inseparable distinction' has now also eroded. The crisis is so serious that scientists are taking a stand, both in the sense that they are becoming advocates, including branding models—which is very clear from the Intergovernmental Panel on Climate Change (IPCC)'s statements—and that they are making pleas to turn research towards urgent knowledge gaps and needs. There is a clear trend towards ceasing to make distinctions between the increasingly inseparable entities of modern science and politics; regarding the latter, it is amazing how the scientific concepts have broken into normal language and policy language with COVID-19. We are facing a reshuffling of evidence and values on both sides of science and policy-making, almost like a 'Hegelian' synthesis of the opposites. However, this 'new synthesis' (if there is one) should not deceive us: we cannot go back to the authoritarian state that decides which values are preferable, and to a science that is dominated by a priori beliefs, the pre-Enlightenment Aristotelian tradition.

Future challenges and further dilemmas

We have enormous challenges in front of us, including climate change and new pandemics. The increase in the frequency of zoonoses (between 1940 and 2004) was first noted in a *Nature* paper that pointed out that 70% of such epidemics were due to agents coming from forests (Ebola, Nipah, SARS, MERS, and now SARS-CoV-2; Jones et al. 2008). The concept of intrusion of the human species into forests and the admixture with wild species is supported by many examples. One of these is the observation that in East Australia the destruction of the natural habitats of bats led the latter to procure their food from urban and peri-urban gardens. But the greatest contribution to zoonoses comes from agriculture. About 40% of the Earth's surface is devoted to food production, and most of it is used for grazing animals and growing their feed. In South America, 70% of the conversion of rainforests between 1990 and 2005 was due to the expansion of new pastures. Animal breeding leads to the emission of large amounts of greenhouse gases, including in particular methane and nitrogen oxides, but also to other important impacts on the planet: use of water in large quantities, eutrophication due to fertilizers, and acidification of soils and waters. In addition, human penetration into forests and climate change have caused two types of modifications in the wild fauna: one is the migration of species in search of food and better habitats, and the second is the disappearance of large-sized mammals in favour of small and middle sizes, including rodents and bats. Both changes have in turn caused an increased risk of spillovers of viruses to human species. Overall, the disappearance of species is happening at a speed that is 100–1000 times quicker than before the Anthropocene.[6]

The answer to future challenges calls for a new philosophical awareness, starting from the issue of responsibility. Since the mid-twentieth century, the philosopher Hans Jonas highlighted a complex and ambivalent relation between technology, causality, and ethics in our age: the progressive increase of our scientific knowledge and technological skills is mirrored by a matching increase in political and moral responsibility (Jonas 1985). Nowadays, we know Jonas's concerns were not baseless: the pandemic and the environmental crisis place twenty-first-century ethics and politics at an increasingly complex level, as the disruption caused to our planet by our economic and technological actions becomes apparent, as does the impact these will have on future generations. The need to forecast and control long-term effects of actions is also just as clear: we cannot afford oversimplifications.

The intellectual pursuit of (over)simplification has deep roots, as it addresses the *moral* need to justify evil and suffering, attributing its responsibility to isolated and detectable causes which would make it easier to act, or at least would appear to make it easier to act. The need for simplification is what determined the success of the war-like metaphors that filled current affairs at the time of COVID-19, but this also seeped into institutional communication that constantly used the image of war against the virus (see also 'Metaphors and science' in chapter 12). The use of war-like metaphors is reassuring, because it conveys the idea of an immediate and clearly pursuable aim that is defeating a deadly enemy. Its use is reductive, as it focuses on immediate causes and the management of emergencies, totally disregarding long-term causes and the need for prevention. As for health and prevention, the emphasis is often on individual responsibility—for instance, stating that members of the community should adopt appropriate behaviours to prevent the spread or mitigate the effects of climate change. Yet, individual responsibility could clash with a perceived disproportionate responsibility, not just because of the cognitive difficulties in appreciating the duration and complexity of causal links, as mentioned earlier in the text. In fact, also from an emotional point of view, distal causes frequently refer to places and people that are very far from our

---

[6] https://www.un.org/sustainabledevelopment/blog/2019/05/nature-decline-unprecedented-report/.

daily experience, making it difficult to feel any form of empathy or moral obligation. Individually, an overarching responsibility could produce an excessively demanding moral rationalism, going against common sense and turning into a diametrically opposite form of de-responsibilization.

In practice, we need to find a balance between individual and collective responsibility, the latter represented by actions such as the Green Deal in Europe and local deliberations. But what about the dilemmas of decision makers? Let us consider an example relevant to the political decisions for the mitigation of climate change and environmental regeneration: organic agriculture. One might think that organic agriculture that does not use synthetic pesticides and fertilizers should be able to partially solve the problem. Unfortunately, organic agriculture has a productivity that is 8–20% lower than conventional agriculture. To feed a growing population, this means that the area devoted to crops needs to be extended, but this clashes with the conservation of biodiversity. In practice, there is no easy choice for policy makers. Though they can certainly claim that their choices are evidence-based, they also need to make their value systems explicit.

And what about the geosphere? The biosphere is in some way regenerable, though in the long run, but the constant extraction of resources, such as minerals, from the geosphere leads to a depletion that is not amendable. Without counting the symbolic loss, at least 10 elements in the geosphere cannot be substituted in human industrial production, comparable to the effect on animal and plant species.

Here are the multiple dilemmas we face: we have to choose between extending cultivated land if we want to promote organic agriculture; we should also increase urban green spaces, for the sake of climate, biodiversity, and human health, but in doing so we would risk further extending the 'ecotone', the admixture of city and countryside that increases the contact with wild species and the risk of zoonoses (particularly in African mega-cities). All these planetary governance problems require good predictions—that is, credible mathematical models that show how the planetary health and each of its components will develop. Just as we have noted in the comments we have made about COVID-19 modelling, we need to be aware of the scientific uncertainties in modelling, but also of the broader picture, which goes beyond science and addresses values. What we ask of politicians is that they have awareness of such uncertainties and the ability to act based on priorities in the middle to long range, something they are not used to. They should realise that we cannot miss the opportunity of stopping climate change, but this would make it impossible to please all their voters. They will need to make choices based on explicit convictions, by establishing a hierarchy among the goals they feel responsible for. The opposite is also the case: by compressing scales of values into the straitjackets of, for example, cost–benefit analyses, everything becomes monetary. Modelling can thus both help and aggravate the present predicaments.

References

Carlston, C.J., G.F. Albery, C. Merow, C.H. Trisos, C.M. Zipfel, E.A. Eskew, K.J. Olival, N. Ross, S. Bansal. 2022. 'Climate Change Increases Cross-Species Viral Transmission Risk'. *Nature, 607,* 555–62, doi: 10.1038/s41586-022-04788-w.

Crutzen, P.J. and E.F. Stoermer. 2000. 'The "Anthropocene"', *Global Change Newsletter, 41,* 17.

Delanty, G. 2020. 'Six Political Philosophies in Search of a Virus: Critical Perspectives on the Coronavirus Pandemic', London School of Economics (LSE), 'Europe in Question' Discussion Paper Series, 156, European Institute, LSE, London.

Descola, P. 2013. *Beyond Nature and Culture,* University of Chicago Press, Chicago.

Gluckman, P. 2014. 'Policy: The Art of Science Advice to Government'. *Nature, 507(7491),* 163–5, doi: 10.1038/507163a.

Grmek, M. 1969. 'Préliminaires d'une Étude Historique des Maladies'. *Ann. E.S.C., 24,* 1473–83.

Guerra, F.M., S. Bolotin, G. Lim, J. Heffernan, S.L. Deeks, Y. Li, and N.S. Crowcroft. 2017. 'The Basic Reproduction Number (R0) of Measles: A Systematic Review'. *Lancet Infectious Diseases, 17,* e420–e428.

Jasanoff, S. 2007. 'Technologies of Humility'. *Nature, 450(7166),* 33, doi: 10.1038/450033a.

Jonas, H. 1985. *The Imperative of Responsibility: In Search of an Ethics for the Technological Age,* University of Chicago Press, Chicago.

Jones, K.E., N.G. Patel, M.A. Levy, A. Storeygard, D. Balk, J.L. Gittleman, and P. Daszak. 2008. 'Global Trends in Emerging Infectious Diseases'. *Nature, 451,* 990–4, doi: 10.1038/nature06536.

Latour, B. 1993. *We Have Never Been Modern,* Harvard University Press, Cambridge, MA.

Locke, J. 1689. *Two Treatises of Government,* 3rd edition, Awnsham and John Churchill, London.

Majone, G. 1989. *Evidence, Argument, and Persuasion in the Policy Process,* Yale University Press, New Haven, CT.

McKenzie, D. 2020. 'The COVID-19 Pandemic Was Predicted – Here's How to Stop the Next One'. *New Scientist,* 16 September 2020. https://www.newscientist.com/article/mg24733001-000-the-covid-19-pandemic-was-predicted-heres-how-to-stop-the-next-one/ (last accessed 2 April 2023).

Meadows, D., D. Meadows, J. Randers, and W. Behrens III. 1972. *The Limits to Growth,* Potomac Associates Books, Washington, DC.

Miller, P. 2022. 'Afterword: Quantifying, Mediating and Intervening: The R Number and the Politics of Health in the Twenty-First Century'. In *The New Politics of Numbers Utopia, Evidence and Democracy,* edited by A. Mennicken and R. Salais, Palgrave Macmillan, London, pp. 465–76.

Morens, D.M. and A.S. Fauci. 2020. 'Emerging Pandemic Diseases: How We Got to COVID-19'. *Cell, 182,* 1077–92.

Morton, T. 2013. *Hyperobjects: Philosophy and Ecology After the End of the World,* University of Minnesota Press, Minneapolis, MN.

Patel, M.K., L. Dumolard, Y. Nedelec, S.V. Sodha, C. Steulet, M. Gacic-Dobo, K. Kretsinger, J. McFarland, P.A. Rota, and J.L. Goodson. 2016. 'Progress Toward Regional Measles Elimination: Worldwide, 2000–2015'. *Morbidity and Mortality Weekly Report, 65,* 1228–33.

Pernice, S., P. Castagno, L. Marcotulli, M.M. Maule, L. Richiardi, G. Moirano, M. Sereno, F. Cordero, and M. Beccuti. 2020. 'Impacts of Reopening Strategies for COVID-19 Epidemic: A Modeling Study in Piedmont Region'. *BMC Infectious Diseases*, 20(1), 798.

Rhodes, T. and K. Lancaster. 2020. 'Mathematical Models as Public Troubles in COVID-19 Infection Control: Following the Numbers'. *Health Sociology Review*, 29(2), 1–18, doi: 10.1080/14461242.2020.1764376.

Saltelli, A., Bammer, G., Bruno, I., Charters, E., Di Fiore, M., Didier, E., Nelson Espeland, W., Kay, J., Lo Piano, S., Mayo, D., Pielke, R. Jr, Portaluri, T., Porter, T.M., Puy, A., Rafols, I., Ravetz, J.R., Reinert, E., Sarewitz, D., Stark, P.B., Stirling, A., van der Sluijs, J., Vineis, P. 2020. 'Five Ways to Ensure That Models Serve Society: A Manifesto'. *Nature*, 582, 482–4.

Waltner-Toews, D., A. Biggeri, B. De Marchi, S. Funtowicz, M. Giampietro, M. O'Connor, J.R. Ravetz, A. Saltelli, and J.P. van der Sluijs. 2020. 'Post-Normal Pandemics: Why COVID-19 Requires a New Approach to Science'. *STEPS Centre Blog*. https://steps-centre.org/blog/postnormal-pandemics-why-COVID-19-requires-a-new-approach-to-science/ (last accessed 2 April 2023).

Weber, M. 2004. *The Vocation Lectures*, edited by D. Owens and T.B. Strong, translated by R. Livingstone, Hackett Publishing Company, Indianapolis, IN.

12

Models as metaphors

Jerry R. Ravetz

Introduction

This chapter discusses philosophical reflections on the intellectual adventure of conducting Integrated Assessment (IA) focus groups with citizens, as presented in this volume.[1] The task of this exercise was ambitious: to bridge the gap between sustainability science and democratic debate in the climate domain. The science component was represented mainly by models, most (although not all) having the appearance of describing future states of the global climate and their consequences for human society. At first it seemed a daunting—indeed, overwhelming—task: it was hard to see how lay participants could meaningfully relate to models whose construction required very special expertise in mathematics and software engineering; and whose comprehension required knowledge of climate science. But having witnessed the debates among the modellers themselves, the research team already knew that IA models are quite problematic products of science. It is freely accepted, even emphasized, among the experts that the models do not provide simple predictions; and so their epistemic status and policy relevance were already open to question. In addition, there was the knowledge that experts are usually 'laypersons' outside their specialities, and that policy makers are generally no more knowledgeable than ordinary citizens. And, in any event, the democratic process involves debate over issues where both expert and lay voices are heard. Hence the IA models were an appropriate vehicle for developing a many-sided dialogue on basic issues.

In the event, the involvement of this 'extended peer community' proved far less difficult than anticipated. For, although the content of the IA models might be arcane, their conclusions were not. Indeed, as predictions of the future state of the planet, the outputs of the models lent themselves to discussion and criticism. This latter arose in two headings: the first, how much the general models could tell us that is relevant to our decisions; and the second, what sort of messages they are.

---

[1] This work was originally published as Ravetz, Jerome R. 2003. "Models as Metaphors." In *Public Participation in Sustainability Science: A Handbook*, edited by Bernd Kasemir, Jill Jager, Carlo C Jaeger, and Matthew T Gardner. Cambridge University Press. Republished with permission of Cambridge University Press.

Jerry R. Ravetz, *Models as metaphors*. In: *The politics of modelling*. Edited by: Andrea Saltelli and Monica Di Fiore, Oxford University Press. © Oxford University Press (2023). DOI: 10.1093/oso/9780198872412.003.0012

In particular, once it was freely admitted that these are not ordinary predictions, categorical statements of what will and what will not be happening, then the lay participants found themselves engaged in an engrossing methodological debate. Given all the scientific talent and technological resources that had gone into the construction of the models, just what use are they if their outputs, whose form is that of categorical statements about the future, are just estimates, or even guesses, shrouded in uncertainties?

The IA focus group procedures developed within our study could actually be an ideal forum in which such issues could be aired. It could well be that in the open setting of these focus groups, free of any of the commitments and prejudices that afflict any policy debate among vested institutional interests, such questions could be framed and expressed all the more clearly, and with greater force. We are not speaking of an 'emperor's clothes' situation, since there has been a continuous and vigorous public debate within the specialist community about the meaning of IA models. But in the context of our research, it was possible for plain people to speak plain words about their confusions and reservations about these scientific instruments. And so the learning experience became universal; they learned about the climate change problems in relation to urban lifestyles, and the experts learned about the different aspects of the usefulness of their tools in the general policy process. In these focus groups conducted with citizens, the IA models—which do not claim to make factual statements or reliable predictions—were seen as useful for enlarging the scope of people's imagination about climate change and the role of individuals in that problem. In that context, models were discovered to be 'poetic'. Stimulated by this experience, this chapter explores the question of whether they could fruitfully be seen as 'metaphors', expressing in an indirect form our presuppositions about the problem and its possible solutions.[2] Although this approach is very different from the traditional understanding of scientific knowledge, it may well be useful in helping science adapt to its new functions in an age of sustainability challenges and scientific uncertainty.

Models as scientific?

For nearly 400 years, our ideal of science has excluded metaphor, for science is supposed to be about exact reasoning, leading to certainty. In scientific discourse and inquiry, the poetic faculty must be tightly constrained, lest it lead us astray. The Royal Society of London expressed it simply (if somewhat obscurely) in its original motto of 1661: 'nullius in verba'. The facts and the power resulting from natural science now serve as the paradigm for all forms of practical knowledge. The 'subjective' studies, or 'gossip', in the words of one distinguished physicist (Ravetz

---

[2] This chapter is to some extent a sequel to earlier reflections on the problems of the proper use of IA models (Ravetz 1997a, 1999). I hope that it will contribute to the sense of a largely successful adventure, or voyage, that our study has given to all its participants.

1971), are deemed inferior, existing only because of the present limits of scientific knowledge and the weaknesses of human intelligence.

It is possible that the triumphs of European science over the past four centuries have been due largely to this attitude of excluding or downgrading the qualitative aspects of experience. Certainly, some confluence of external and internal factors made possible the unique rise of our science, to achieve a degree of knowledge and power that could scarcely have been dreamed of in previous civilizations. But now, at the start of the twenty-first century, we realize that the problems of 'the environment' are becoming challenges to the 'sustainability', or survival, of our civilization; and that our 'urban lifestyles', made possible by our enveloping science-based technology, are largely responsible for this perilous situation. Our simplistic, myopic technology threatens to destroy our own habitat, and our reductionist science, by definition incapable of grasping systems thinking, is inadequate for managing the tasks of cleanup and survival. The quantitative social sciences that are designed around the imitation of Victorian physics become ever more clearly seen as caricatures. Instead of being genuinely scientific in the way that their practitioners so ardently desire, such disciplines are merely 'scientistic', misconceived parodies of real knowledge. The modern programme of scientists 'teaching truth to power', deducing correct policies from incontrovertible facts, is, in the environmental field, in tatters. We now have an inversion of the classic distinction between the hard objective facts produced by science and the soft subjective values that influence policy. Now we have decisions to make that are hard in every sense, for which the scientific inputs are irremediably soft. The goal of the whole enterprise, 'sustainability', is now recognized as something other than a simple, scientifically specifiable state. Rather, sustainability (especially when the social and moral elements are included) is something of an 'essentially contested' concept in the sense of Gallie (1955). In these circumstances the denial of rhetoric, as part of the traditional reduction of reality to quantitative attributes, can now be seen as a profound metaphysical prejudice, one of the patently counter-productive elements of our intellectual heritage.

All these problems are exposed most sharply in the use and interpretation of environmental models, particularly IA models that are cast in the form of describing future states of the global climate and their consequences for society. These 'models' are themselves of an unusual sort. They are not miniature globes with water and clouds swishing around, and in fact they cannot even be seen. Rather, they are sets of mathematical relationships, embodied in computer programs, which (it is intended and hoped) simulate some of the interactions in the bio-geosphere. To understand how any model works is indeed a task for experts; but to ask what it tells us, indeed why it has been constructed, is open to any interested citizen.

The outputs of the IA models are represented as assertions concerning quantitative indicators, expressed at some future time. But they are not simply 'predictions' in the sense of the classical philosophy of science. In our traditional understanding, the statements of natural science are intended to be tested against experience, to determine the truth or falsity of the theory from which they are deduced. However, a computer model is not a 'theory' in the sense of the achievements of Newton and Einstein. The models are not expected to be 'true' or 'false' in the classic scientific sense. Further, in the case of most IA models, the outputs relate to times in the future which are too far away for any practicable 'testing'. And in any event, practitioners now agree that even if one were to wait for the requisite number of years or decades, the actual states of those indicators would most probably be quite different from those 'predicted' back in our present. If seen from the perspective of classical philosophy of science, all our models are trivially 'false'.

Those who develop and use these models must then become creative methodologists. The models are said to have a variety of heuristic functions. Prominent among these is clarifying our understanding of our assumptions. But it is not clear whether this relates to the assumptions made by society in general or by policy makers about the environment, or merely to the assumptions made by the modellers about the structure and inputs of their models. Models also might provide some indications of the way things will turn out in the real world; but then again, they might not. All this endeavour with models is very important, as our assessment of the future is genuinely quite crucial in the setting of public policy. But IA models are clearly seen to be lacking in a methodological foundation in the successful practice of natural science. Therefore, the justifications of this sort of modelling will not be able to succeed within the framework of the traditional conceptions of scientific knowledge and practice. They are based less on a successful practice of advancing knowledge, and increasingly on an embattled faith.

If these models are not fully 'scientific', but to some significant extent merely 'scientistic', how are they to be understood as objects of knowledge? Under the challenging conditions of our focus groups, fresh insights could be generated which would not be likely to emerge within the groves of academe. Thus, in one regional case study the discussions gave rise to the notion that climate change models are a form of 'seduction' (Shackley and Darier 1998). The term is used in a nonsexual sense following the French philosopher Baudrillard, as 'a game with its own rules', and refers rather more to the modellers than to the models themselves. Reviewing the variety and confusion among the explanations of the uncertainties, dependence on special assumptions, and value-loading of the models, they eventually arrive at the explanatory formula: 'truths/untruths'. For the 'IAMs [Integrated Assessment Models] are hybrids ... a mixture of conditionally valid uncertain and indeterminate knowledge—data, theory and insights—combined together in a fashion which generates further indeterminacy'. Further, 'several types of ignorance [are] involved, related to processes, phenomena and data'.

The authors argue that the advocates of the models alternate between using strong and weak claims for them, in order first to recruit possible supporters, and then to keep them on board when the inadequacy of the models becomes apparent. This is what is understood as 'seduction'; but it should be observed that the process may well be directed even more at the modellers themselves, to maintain their own sense of worth in the face of disillusioning experience. Such an explanation was offered by Brian Wynne in his classic study of the International Institute for Applied Systems Analysis (IIASA) energy model, whose designers, aware at one level of its quite crippling flaws, had to practise a sort of Orwellian doublethink on themselves as well as their patrons (Wynne 1984).

In historical perspective, the dilemma of the modellers is really quite ancient. The policy-related sciences have been in such a bind for a very long time. The earliest mathematical social science, astrology, struggled constantly with suspicious clients and with methodological conundrums (as the 'simultaneous births, diverging fates' paradox). The first applied natural science, medicine, faced harsh criticism and even derision until quite recently. It is a sobering thought that academic medicine, one of the mainstays of the university curriculum for well over half a millennium, was, in retrospect, absolute nonsense. Those patients whose condition could be treated by an 'empiric' had a chance of effective treatment; those who went to the learned doctors of physic needed to trust their luck, right through to Victorian times. It was only the triumphs of the applications of science in the present century that made success the natural and expected condition for science. Only recently has research science been accepted as the paradigmatic form of genuine and effective knowledge.

Now the sense of the power and limits of science is changing quite rapidly. We know that science can fail to produce a desired good in the form of safety and wellbeing; and we also know that science can produce evil, accidentally or intentionally. The authors of the seduction analysis suggest that we now go beyond 'enlightenment rationality and instrumentalism, and to open up discussion to the messy processes of thinking, creating and imagining that we all engage in through everyday practice'.

Metaphors and science

In that spirit, we feel justified in searching further afield, seeking other conceptions of knowing, with which we can explain and guide our actions in this still significant practice. Let us consider 'metaphor', in spite of the long tradition within science of denying and deriding metaphor as an inferior form of expressing knowledge (for a note on the terms 'model' and 'metaphor', see Box 12.1).

Box 12.1 Models and metaphors

'Model' is a word with many meanings. In this context, it refers to computer programs designed to mimic the behaviour of particular complex systems. Models are used in the cases where neither theoretical understanding, nor experimental verification, nor statistical analysis is available in sufficient strength. The variables in the model represent observable properties of the system, and the structure of the model represents the relations among them that are known and also capable of simulation. Models normally require 'adjustment' elements introduced ad hoc to make their outputs plausible—and validation of the models is always indirect.

'Metaphor' is a rhetorical device, meaning 'carrying beyond'. It refers to the denotation of an idea by a term which literally refers to something else. Its explicit rhetorical use, as in poetry, is to add dimensions of meaning beyond those available in prose. Although the practice of science is believed to be antithetical to poetry, any process of naming, particularly of new theoretical entities, relates them back to other ideas and in that sense is metaphorical. In computer models, the metaphors conveying extra dimensions of meaning tend to be hidden, both in the general assumptions about the world that make models relevant, and also in the particular assumptions about reality and value that shape the model and its outcomes.

Surprisingly, we find metaphor embedded in the most apparently scientific sorts of discourse. Darwin's theory of evolution relied explicitly on metaphors, such as the 'Tree of Life'; and what is 'Natural Selection' but the greatest metaphor of them all? Even the concept of 'species', so central to Darwinian theory, has functioned as a metaphor for discreteness, fixity, and indeed purity, so that biologists are only now discovering how much they have been biased against recognizing the importance of hybrids (Brooks 1999).

Even in physics, the very structure of basic theories, like thermodynamics, is conditioned by the metaphors embedded in it, such as 'ideal heat engine' and 'efficiency'; and these reflect the perspectives and values of the society in which the theories were forged (Funtowicz and Ravetz 1997). In contemporary policy-relevant sciences, the importation of social values is clear in such titles as 'the selfish gene'. Overarching metaphors like 'growth' are translated into particular sorts of social-scientific language, and are then given very particular sorts of policy implications. One may then legitimately inquire about the extent to which the practices in such fields, in their criteria of value and adequacy, are themselves influenced by the same social values that provide their metaphors (Lakoff and Johnson 1980, Luks 1999).

Thus, we find that in spite of our pretensions to managing so many aspects of our affairs scientifically, that practice is both described and informed by metaphors, themselves embodying societal and cultural values which doubtless shape the practice itself. Reflecting on this state of affairs, we can welcome the prevalence of metaphors, but we can also regret the absence, hitherto, of awareness of their prevalence. For without awareness of our driving metaphors, our supposedly scientific practice is afflicted by a sort of false consciousness of itself. Earlier theorists ascribed this particular defect to other sorts of knowing, assuming that science, by definition, is immune to it. But now we see that such confidence was misplaced. And a practice governed by self-delusion is vulnerable to every sort of distortion and perversion. Recall that we are not talking about the rock-solid experimental sciences of yesteryear, but about the sciences which are both intimately related to policy and also necessarily uncertain because of the complexity of their assigned tasks.

If we then propose to embrace metaphor as an explanation of a practice, in this case the mathematical methods used in environmental analysis, then our conceptions of its objects, methods, and social functions (three related aspects) must come up for review, along with the reframing of the appropriate criteria of adequacy and value. This is a large task, to be conducted through dialogue among all those concerned with the problem. The present remarks are intended only to show why such a dialogue is legitimate and indeed necessary, and to indicate the sorts of theoretical lines along which it should proceed.

We may start with Michael Thompson's (1998) insight that, while the future is unknowable, this area of ignorance is to be viewed positively, as an opportunity for the growth of awareness through a dialectical process. The core of this awareness is that of our ignorance. In that way, studies of the future can induct us into Socrates' philosophical programme, wherein we become aware of our ignorance. Since the whole thrust of Western philosophy since Descartes has been to control, deny, and ultimately conceal our ignorance, this is a radical programme indeed (Ravetz 1997b).

We are thus confronted with two conceptions of the task of using models. One is based on the faith that scientific methods can be extended to knowing the future, and hence to bringing it under control. This conception is expressed in what I have called the 'elite folk sciences' of reductionist quantification of the natural and human realms alike (Ravetz 1994/5). These include the so-called 'decision sciences', along with mainstream economics and the predictive computer modelling fields. Their language is rich with metaphors, but they are all taken from the 'possessive individualist' conception of humanity and nature. Their methodology is an imitation of the 'hard' natural sciences, and they attempt to operate hegemonically in all the relevant fields. Needless to say, ignorance is a severe embarrassment to such sciences, as it presents itself as a simple refutation of their claims of total knowledge and complete control.

The other conception embraces uncertainty and ignorance, and welcomes the clash of distinct perspectives. Its style is dialectical, recognizing that the achievement of final truth is a false and misleading goal. This conception is permitted only at the very margins of the practice of 'matured' science, as in discussions within small colleague communities at open research frontiers. Otherwise, in the pedagogy and popularization of research science, certainty rules. Up to now, that particular conception of science, inherited from a triumphant and triumphalist past, has been successful in that it has provided many more 'goods' than 'bads' for humanity (or at least its 'fortunate fifth'). Hence it cannot be refuted in its own terms. But now that our science-based technology reveals its negative impacts on the natural environment and on ourselves as well, and in ways that cannot be controlled, predicted, or even anticipated, the assurance of triumph is weakened. Uncertainty at the policy interface is now recognized, however reluctantly, as inescapable. Hence science as a whole, and especially the 'predictive' fields, must now embrace real uncertainty, or sink into undeniable confusion and vacuity.

The dilemma can be seen creatively, if we understand these predictive sciences as telling us less about the natural world of the future, and more about our social and intellectual world of the present. In that sense, we see them as metaphors, providing knowledge not by mere straightforward assertion, but rather by suggestion, implicit as well as explicit. In that way the loss of the pretence of scientific certainty can be seen as a liberation, whereby our discourse about ourselves in nature is opened to the enhanced understandings of metaphor and poetry. This vision is amply borne out by the experience of our focus groups, where the confusion caused by the scientistic understanding of the models gave way to the creativity of discourse about their metaphorical meanings. Disagreement was thereby freed from its negative interpretation, and could then be appreciated as the expression of complementary visions of a complex reality.

Metaphors in environmental modelling

But how are we to find metaphors in the forbidding, frequently impenetrable thicket of formulae and computer codes that constitute mathematical 'models' of environmental processes? We can be sure that they will not be patent, announcing themselves as transferring meaning from some other term to the one under scrutiny. So we look for implicit features of the construct, of which perhaps even the modellers themselves may not have been aware. One place is in the assumptions, cast in mathematical form but expressing values as cogently as any cri de coeur. For example, whenever we put a numerical value on things that will exist at some future time, we are assuming a particular 'social discount rate'. This expresses the price that we, as individuals or a society, are willing to pay for deferring our use (or consumption) of resources. A high 'social discount rate' means that future rewards have little value; it expresses the philosophy 'What's posterity done for me?'; conversely, a low social discount rate stands for 'We are here as stewards of our inheritance.' In between, the choice is driven by values and politics, as filtered by fashions in the relevant expert communities. Yet policy conclusions, apparently the outcome of rigorous theoretical logic realized in precise mathematical calculation, can depend quite critically on the size of this concealed quantified value commitment. In the design of permanent structures, either in the public and private sectors, the assumed working life (and hence the quality of construction) will depend quite critically on this assumed discount rate. And in the evaluation of environmental goods, the 'present value' of an ongoing resource depends critically on the rate at which it is discounted into the future.

It can be quite instructive to witness IA modellers displaying the predictions and recommendations of their model to three-digit precision, and then to learn that they have never tested it for sensitivity to the assumed social discount rate. There is a simple relation between decrease in value and discount rate; for example, we may define the 'throwaway time' of an object as the length of time required for it to be reduced to a tenth of its present value. This is equal (in years) to 230 divided by the discount rate. So for a 10% discount rate, we throw away before a 'generation' of 25 years has elapsed; with 7%, the throwaway time is 33 years. If we really value the wellbeing of our grandchildren, and have 'throwaway time' of, say, 60 years, then our social discount rate is only 4%. The future that we construct through the choice of discount rate is (in this respect) a metaphor for our conception of the good life in the here and now, as revealed through our evaluation of the future.

Another implicit metaphor lies in the choice of the attribute of the future which is to be salient for our scenarios. There are too many complexities and uncertainties for anything like 'the whole picture' to be conveyed at once. So modellers necessarily choose some aspect, which will involve a design compromise between what is scientifically reasonable, and what is humanly and politically meaningful. The early focus on increase of 'global mean temperature' related the debate to the capabilities of the leading models, and it also cohered with the comforting assumption that change in the future would be smooth and somehow manageable. It was left to commentators to elaborate on the implications of temperature rise, which included melting ice-caps, changing crop patterns, and new diseases. More recently, we have become more aware of instabilities and extreme phenomena occurring on a regional or local scale; the vision is now becoming more 'catastrophist'. These irregular phenomena are much less amenable to scientific treatment, but they offer convenient 'confirmations' of climate change whenever the weather comes up for discussion. Neither focus is 'right' or 'wrong' in any absolute sense; each has its function, dependent on the context of the debate at any time. Each is a metaphor for a predicament which defies control and perhaps even defies full understanding.

In analysing these metaphors, I am suggesting a sort of 'deconstruction', but not in a negatively postmodern spirit of demystification or debunking. Nor am I asserting that environmental models are simply, or nothing but, metaphors. Rather, as in the analysis of works of creative art, we can use the idea of style to illuminate what the work is about. This is expressed in the less declaratory aspects of the work, but provides a key to its context, framework of ideas, paradigm, perspective, or *Weltanschauung*—call it what you will. In the visual arts, the 'style' relates less to the explicit theme or subject of the production, and more to silent choices made by the creator, on technical aspects of the work or on particular implicit thematic materials. In written work, style can relate to vocabulary, diction, place of the narrator with respect of story and reader, and so on. 'Style' is used by scholars to place works (sometimes quite precisely) within ongoing traditions and, alternatively, to tease out deeper layers of meaning.

Among computer models, stylistic differences relevant to users are most easily discerned in their outputs. Even the choice between digital and graphical displays reflects cosmological metaphors. This is expressed in Oriental philosophy between the Yang and the Yin, or in classic computer terms between the IBM of the New York corporation and the Apple of the California garage. Digits provide information that is precise (perhaps hyper-precise) on details, but they fail to convey any sense of overall shape. Graphs are more expressive, but are vague; and a collection of curves all climbing upwards at roughly the same rate does not stimulate either the eye or the mind. The representation of uncertainty is even more fraught. To accompany each principal curve with others that display 'confidence limits' can give a seriously misleading impression of the precision and information content of those supplementary curves. It might be best (or rather least worst) to show curves as 'caterpillars', consisting of shaded areas with gradations from centre to edges; but I have seen hardly any examples of this.

If we adopt maps for our display, we may be involved in another level of inference and interpolation (from the global to the regional for the case of IA models on climate change and its relations to social activities). We also incur a significant risk of conveying a false verisimilitude to users. The partly metaphorical character of maps has not been generally concealed (unlike in the case of numbers). For example, the use of standard colours and conventional symbols is obvious (Funtowicz and Ravetz 1990). When a map makes a patent distortion of what is on the ground, like the graph maps of the London Underground tradition, the metaphor is clear. Even there, it has been discovered that people (and not only tourists!) sometimes orient themselves solely by the Underground map, ignoring coincidences on the ground that could not be conveyed on the plan. (The most famous of these is the pair of stations Bayswater and Queensway, within a stone's throw of each other in reality and yet unrelated on the map.)

Indeed, we may say that the more realistic and powerful the map display becomes, the more urgent the question of its epistemic character becomes. It is all too easy for maps to become instruments of seduction, where the display is taken for the reality. This can happen even when the given output is only one among several alternatives. Somehow each alternative 'vision' gets a quasi-real status: if this and that about the general picture are as we assume, then the detailed future will be just as shown on the screen. The task for users may then seem to become one of making 'correct' or 'the best' assumptions, to predict the 'real' future state. In spite of all the variety and uncertainty that may be built into the model and clearly expressed in the instructions for its use, the combination of deductive structure and compelling display can tend to make its interpretation fatally scientistic after all. Those who use them extensively are at risk of becoming seduced by the assumptions concerning reality and value that are embedded in their structures (Dyson 1999).

There are various ways to guard against such a development. One is for the models to be able to convey the 'bad' along with the 'good' news. Particularly when regional models are employed, anything that shows the effects of general constraints (as in land or water supply), or contradictions between various goals, is to be welcomed. When they cause discomfort by showing unexpected or unwanted consequences of a position assumed to be good and natural, they can have real educational benefits. Better still, if they are employed within a dialogue, so that the experiences of disappointment and disillusion are shared, their Socratic function is enhanced. For then it becomes public knowledge that just as men formerly made gods in their own images, now they construct apparently scientific futures out of their hopes and dreams based on the past.

When embedded in a dialogue full of surprise and shock, the model can then function truly as a metaphor. It is recognized as carrying not a literal truth, but an illumination. And what the bare output (numerical, graphic, or cartographical) lacks in enhancing an aesthetic imagination, it can compensate for in its development of our self-awareness. Knowledge of our ignorance is, it was said long ago, the beginning, or rather the prerequisite, of wisdom. This sort of knowledge has been systematically excluded from our intellectual culture for the past 400 years; and it could be that this ignorance of ignorance, and the scientific hubris to which it gives rise, is responsible, in no small measure, for the present perilous character of our total scientific system. While in some ways this is still plausibly in the image of the conqueror of Nature, even more does it remind us of the Sorcerer's Apprentice.

Moreover, the recognition of models as metaphors will have a profoundly subversive and liberating effect on our very conception of science itself. For a metaphor has a reflexive, even ironic character (O'Connor 1999). When I say, 'My love is like a red, red rose', I know as well as the reader that this is literally false. There is a deeper truth, which may be explicated later in my story, but starting with the metaphor, I draw the reader into a little conspiracy, perhaps in its way a seduction. We share the knowledge that I have said something apparently false and ridiculous, and we will now play a game where I show how this apparent non-knowledge actually becomes a better knowledge than a photographic description. Of course, they may not want to play; some people find metaphors peculiar, useless, and distasteful.

Healing the amputation of awareness in science

There is a long history in Europe, partly cultural and partly political, of a reaction against metaphor. One can find it in the early Protestant reformers, who wanted the Bible to be understood as a plain history for plain men. This, they thought, would eliminate the corrupted spiritual expertise of the priesthood. Later, the same impulse was realized in the forging of a new conception of science. The prophets of the Scientific Revolution all put literature in a separate, and usually unequal, category from the knowledge derived in the experimental-mathematical way. Subsequent spokesmen of science stressed the need for clear thinking, for a hobbling of the imagination, lest the mind be led astray.

But now we find ourselves in a peculiar situation. Beneath the hard surface of scientific discourse, metaphors abound; we have 'chaos' and 'catastrophe' in mathematics, and 'charm' in physics. It is just possible that those who create and popularize these metaphors are unaware of their complex character. They might think of them as cute and suggestive names, rather than as conveying realms of knowledge and presuppositions, which can be all the more powerful for being unselfconscious. In this way, the creative scientists are, all unawares, poets *manqués*; in their nomenclatural practice they violate the principles on which their knowledge is claimed to rest.

Were this a matter concerning only theoretical scientists ensconced in academe, it would be of purely 'academic' interest. But the pretensions of science, embedded in and unselfconsciously conveyed by the apparently impersonal conceptual instruments of analysis, can, in cases of environmental policy, deceive and confuse the scientists as much as their audience. The numbers, graphs, and maps announce themselves as objective, impersonal facts; they are a whole world away from the red, red rose of the poet. Appreciation of them as metaphors requires an even greater sensitivity than that needed to grasp that the red, red rose is not a photographic description. But it is all the more difficult because of the amputation of awareness that is instilled by the standard education in science.

Students of the traditional sciences are still formed by a curriculum which teaches by example that for every problem there is just one and only one correct answer; and the teaching is reinforced by the discipline of the exam. Having been thus force-fed for a decade or more until they emerge as 'Doctors of Philosophy', students or researchers generally do not know what they are missing of the total picture of scientific knowledge in its context. Those who know about this amputated awareness do not necessarily think it is a bad thing. Indeed, it was argued, by Kuhn himself, that it is integral to the process of science and essential for its success (Kuhn 1962). The deficiencies and dangers of 'puzzle-solving' in 'normal science' become manifest only when that process fails in its own terms: when, in Kuhn's words, there is a 'crisis'. This can occur in research science as a precursor to a Kuhnian 'scientific revolution'. More commonly, now, it occurs whenever science is in a 'post-normal' situation where facts are uncertain, values in dispute, stakes high, and decisions urgent. Then puzzle-solving is at best irrelevant and at worst a diversion.

The amputated awareness of science is directly challenged when models are introduced as a means of education about global environmental problems and urban lifestyles. The experience of our focus groups has repeatedly shown that the 'plain man's question', namely, 'Are these real predictions?' will, if accepted at its own level, produce nothing but confusion. If the models are claimed to predict scientifically, refutation follows swiftly; but if they are not predictors, then what on earth are they? The models can be rescued only by being explained as having a metaphorical function, designed to teach us about ourselves and our perspectives under the guise of describing or predicting the future states of the planet. In that process, we all have the opportunity to learn not merely about ourselves, but about science as well.

In all this, there is another lesson to learn about science. The triumphs of science and the technology based on it are undeniable; no one could, except provocatively or mischievously, say that science is nothing but metaphor. But that same science has produced our present predicament, where our urban lifestyles are clearly unsustainable. How is that total system of knowledge and power to be transformed, so that it does not destroy us in the end? Can the understanding which has become the leading problem for civilization simply turn around and constitute the total solution? My argument, supported by the ULYSSES[3] experience, is that a new understanding of the science that is employed in the policy processes is necessary, and that the awareness of metaphor has its part to play.

Conclusion: ULYSSES and the future

There is a consensus among its participants that this enriched understanding has been an important achievement of the present study. Through the interaction of experts with laypersons, we have found that the question for discussion and mutual learning is not restricted to the properties of this or that particular

[3] ULYSSES, for Urban Lifestyles, Sustainability, and Integrated Environmental Assessment, is a collaborative project funded by the European Commission under the Fourth RTD Framework Program, Environment and Climate, Human Dimensions of Climate Change; contract ENV4-CT96–0212, 1996–1999).

model in relation to its user friendliness. In the many interactions with intelligent and thoughtful laypersons, these methodological issues have come up repeatedly. While related considerations have pervaded the whole work of the present study, the 'metaphor' metaphor (so to speak) has even been introduced explicitly in one of the regional case studies. Here, the participatory process has been portrayed as a 'voyage of discovery' on which one of the passengers is a somewhat odd 'Mr Computer' (De Marchi et al. 1998). And, crucially, all participants have learned important lessons, the experts no less than the laypersons. Such an outcome, that in these conditions the experts also have something important to learn, is the essence of post-normal science. They can teach the laypersons something from their expertise, but the others can teach them something about that expertise itself. When all sides are aware of their mutual learning, and all sides gain thereby in self-understanding and mutual respect, then (and only then) can there develop that element of trust among participants, which is becoming recognized as the essential element in our making progress towards a sustainable world. In this way, the experience of the study discussed in this book, where (perhaps uniquely) the products of leading-edge integrated assessments of climate issues were exposed to scrutiny and earnest criticism by the supposed beneficiaries of the research, provides lessons of the utmost importance for us all.

A general recognition of models as metaphors will not come easily. As metaphors, computer models are too subtle, both in their form and in their content, for easy detection. And those who created them may well have been prevented, by all the institutional and cultural circumstances of the work, from being aware of their essential character. Furthermore, the investments in the scientistic conception of models, in the personal, institutional, and ideological dimensions, are enormous and still very powerful. But the myth of the reductionist-scientific character of our studies of the future, and indeed of all complex systems, cannot hold. Only by being aware of our metaphors, and our ignorance, can we fashion the scientific tools we need for guiding our steps into the future, now appreciated as unknown and unknowable, but where our greatest challenge lies. Real, working dialogues between the community of experts and the 'extended peer community' are essential in improving this awareness on both sides. The initiative of the study discussed in this volume has illuminated the problem of the use of science for policy, and it has shown the ways towards its solution, through dialogue, learning, and awareness. In this, I hope that the understanding of models as metaphors has played a part.

References

Brooks, M. 1999. 'Live and Let Live'. *New Scientist*, 3 July 1999, pp. 33–6.
De Marchi, B., S.O. Funtowicz, C. Gough, A. Guimarães Pereira, and E. Rota. 1998. *The ULYSSES Voyage: The ULYSSES Project at the JRC*, EUR 17760EN. Joint Research Centre, European Commission, Ispra.

Dyson, E. 1999. 'Losing Touch with Reality'. *The Guardian*, 4 March 1999, Online Supplement, p. 11, London.

Funtowicz, S.O. and J.R. Ravetz. 1990. *Uncertainty and Quality in Science for Policy*, Kluwer, Dordrecht.

Funtowicz, S.O. and J.R. Ravetz. 1997. 'The Poetry of Thermodynamics'. *Futures*, 29(9), 791–810.

Gallie, W.B. 1955. 'Essentially Contested Concepts'. *Proceedings of the Aristotelian Society*, 56, 167–98.

Kuhn, T.S. 1962. *The Structure of Scientific Revolutions*, University of Chicago Press, Chicago, IL.

Lakoff, G. and M. Johnson. 1980. *Metaphors We Live By*, University of Chicago Press, Chicago, IL.

Luks, F. 1999. 'Post-Normal Science and the Rhetoric of Inquiry: Deconstructing Normal Science?' *Futures*, 31(i), 705–19.

O'Connor, M. 1999. 'Dialogue and Debate in a Post-Normal Practice of Science'. *Futures*, 31, 671–87.

Ravetz, J.R. 1971. *Scientific Knowledge and its Social Problems*, Oxford University Press, Oxford.

Ravetz, J.R. 1994/5. 'Economics as an Elite Folk-Science: The Suppression of Uncertainty'. *The Journal of Post-Keynesian Economics*, 17(2), 165–84.

Ravetz, J.R. 1997a. *Integrated Environmental Assessment Forum: Developing Guidelines for 'Good Practice'*, ULYSSES WP-97-1, Center for Interdisciplinary Studies in Technology, Darmstadt University of Technology, Darmstadt.

Ravetz, J.R. 1997b. 'A Leap into the Unknown'. *The Times Higher*, 28 May 1997, London.

Ravetz, J.R. 1999. 'Developing Principles of Good Practice in Integrated Environmental Assessment'. *International Journal of Environment and Pollution*, 11(3), 1–23.

Shackley, S. and É. Darier. 1998. 'Seduction of the Sirens: Global Climate Change and Modelling'. *Science and Public Policy*, 25(5), 313–25.

Thompson, M. 1998. 'Three Visions of the Future'. Paper presented at the kick-off workshop of the VISION project, 6–7 May 1998, Maastricht.

Wynne, B. 1984. 'The Institutional Context of Science, Models, and Policy: The IIASA Energy Study'. *Policy Sciences*, 17(3), 277–319.

Epilogue: Those special models

A political economy of mathematical modelling

Andrea Saltelli and Monica Di Fiore

Introduction

As curators of this volume, we hope we don't do violence to the contribution of our authors by suggesting that all share a perception of the 'state of exception' enjoyed by mathematical models. To put it another way, mathematical models are special. In what sense? We try here to chart what is exceptional.

Mathematical models can dispose of a possibly unlimited repertoire of methods from mathematics, from calculus to operational research. No other instance of quantification—not even statistics—can draw from a comparable palette. Their technical versatility allows models an unending colonization of knowledge: there is hardly an avenue of life or nature that is not permeated, observed, regulated, or managed by some kind of model. According to Steven Shapin, we should be grateful for what he calls 'invisible science', which is omnipresent and improves our lives (Shapin 2016). By the same token, one can argue that we should be grateful to 'invisible models' (Saltelli 2019b).

A second exceptional feature of mathematical modelling is that, unlike statistics, it is not a discipline in itself, and can be seen in action in an archipelago of disciplines and practices.

A third surprising peculiarity of mathematical modelling is that it largely escapes the lenses of sociology of quantification. This latter is often concerned with statistical indicators, algorithms, cost–benefit analyses, even multi-criteria analysis, or with quantifications that necessarily incorporate modelling in their construction—again, algorithms and artificial intelligence—but rarely in what would be taken by practitioners to be the core business of mathematical modelling, constructs involving calculus or one or another of the many engineering approaches made possible by mathematics and computers.

A fourth—but surely not less important—feature of mathematical models is that they enjoy the prestige and epistemic authority accruing to mathematics. Hardly any other form of quantification, not even one based on statistics—can make similar claims on truth telling. Once a system of differential equations is solved to

Andrea Saltelli and Monica Di Fiore, *Epilogue: Those special models*. In: *The politics of modelling.* Edited by: Andrea Saltelli and Monica Di Fiore, Oxford University Press. © Oxford University Press (2023).
DOI: 10.1093/oso/9780198872412.003.0013

simulate a system, we have the simulation delivered together with something that is simultaneously a tool, an interpretation, and a representation of the system (Morgan and Morrison 1999).

What are the consequences of this special status of models?

Relatively shielded by the inquisitive gaze of sociologists, mathematical modelling may more easily advance pretences of neutrality. Niklas Luhmann is probably right that the role of science in policy is to 'deparadoxify' political decisions by making them appear to be descended from science (see discussion in King and Schüt 1994). In both the US and the UK there was considerable debate when COVID-19 decisions were instrumentally attributed to following the science (Pielke 2020, Rhodes and Lancaster 2020). Model is the perfect political instrument of deparadoxification.

A second consequence of the status of modelling is a lack of standardized procedures for quality (Eker et al. 2018, Padilla et al. 2018). It is a discipline without recognized leaders, authoritative fora, or 'antibodies' (Saltelli 2019a) to effectively fight possible abuses or degeneration. For Jerome R. Ravetz (1971),

> [W]hile gross errors in technique can be detected by the discovery of pitfalls in published work, the more subtle components of method are not easily accessible to anyone who has not already invested a part of his life in working in the field. Thus, reforming a diseased field, or arresting the incipient decline of a healthy one, is a task of great delicacy. It requires a sense of integrity, and a commitment to good work, among a significant section of the members of the field; and committed leaders with scientific ability and political skill.

This quote—when applied to mathematical modelling—frames well a key aspect discussed in this book. The predicaments of the current situation, as many of our authors argue, lie in this omnipresence of models; they are capable of delivering results that range from the extremely useful to the grossly harmful, in a situation characterized by knowledge asymmetry between developers and users. This status of things poses a clear governance problem.

In the absence of 'a commitment to good work among a significant section of the members of the field and committed leaders with scientific ability and political skill', the extraordinary versatility of models as both instruments with which to do things and representations to reveal things (Morgan and Morrison 1999), is the reason for their vulnerability.

The 'ritual' use of the methods, professional opinions, and symbols of mathematical modelling too easily obfuscates the difference between model and world, at times suppressing alternative understandings under those perspectives that are left out (see 'Mind the unknowns').

Models may offer the perfect ground for trans-science (Weinberg 1972)—that is, for practices that may be formulated in the language of science but that cannot be solved by it. In principle, a model can be written that holds all the infinite features of the system, like Borges' one-to-one map of the Empire (Borges 1946), but such a model is useless, as its predictions are correspondingly uncertain (see 'Mind the hubris'). In statistics, some have noted the possibly ritualistic use of methods (Gigerenzer 2018)—the same happens in models, but again less attention is paid to this.

The lesson from post-normal science, discussed in the present volume (see 'Sensitivity auditing'), is that more is needed to 'defog the mathematics of uncertainty' (Funtowicz and Ravetz 1990) and to keep under control pseudo-science, where results appear so precise as to promise accuracy, while instead this is only achieved because important uncertainties have been neglected.

The political economy of mathematical modelling

Reading the contribution of our authors, one might be tempted to identify a specific political economy of mathematical modelling (PEMM). A feature of this PEMM is that size and complexity are related to epistemic authority: the larger the model, the more professional its authors. This may lead to an instrumental use of complexity. We borrow again from Jerome R. Ravetz to suggest that this might be an important cause of corruption of scientific practice:

> [w]e have perhaps reached a complex epistemic state, where on the one hand 'everybody knows' that some numbers are pseudo-precise and that numbers can be gamed, while the game works only because most people don't know about it.
>
> (Ravetz, pers. comm.)

A gist of the reflection of Morgan and Morrison (1999) on 'models as mediators' is that models are partly independent from both the theory and the world they represent: 'models are not situated in the middle of an hierarchical structure between theory and the world' (p. 17). Their versatility enables them to act as mediators in a vast array of tasks.

More is needed in the construction of models than is in either theory or the world, including elements of craft skill, as noted by system biologist Robert Rosen (1991), who opened a new dimension in the meaning of models by noting that reflexive anticipatory systems—those that can adapt to changes yet to occur—are those that possess a model of themselves (Louie 2010).

Models—especially the large ones—are not made solely of mathematics. They are wedded, often inextricably—to large sets of data. Edwards (1999) describes this relation as 'incestuous' to make the point that, in modelling studies, using data to prove a model wrong is not straightforward. Thus, models do not meet the criterion for genuine science (as distinct from pseudo-science) that was established by the philosopher Karl Popper, that of being capable of refutation. Since, in addition, they are not verified by either induction or deduction (the traditional criteria), their scientific status remains somewhat problematic.

Edwards speaks of 'data-laden models, model-filtered data', and this may worsen the knowledge asymmetries between models users and developers often signalled by practitioners (Jakeman, Letcher, and Norton 2006). The considerations of Edwards (1999) are very similar to those made by historian Naomi Oreskes (2000). Since models are a 'complex amalgam of theoretical and phenomenological laws (and the governing equations and algorithms that represent them), empirical input parameters, and a model conceptualization', when a model prediction fails, what part of all this construct was falsified? The underlying theory? The calibration data? The system's formalization or choice of boundaries (Oreskes 2000: 35–6)?

An important characteristic of the PEMM is that among the various scientific communities, modellers appear to be the ones which, with hardly any contestation, have the strongest grip on policy. As discussed in our Preface, there are crucial policy settings—climate change being one—where models enjoy a quasi-uncontested role. Models can calculate the social cost of carbon dioxide averaged over the next three centuries with high precision, and the resulting numbers are taken as a meaningful input for policy formulation, however paradoxical this might appear to several of the authors of the present work.

To take another example, economic models whose realism is fiercely contested do continue to play an important role in the formulation of policies (see 'Mind the consequences'). Because of the exceptional status of models, they are only exceptionally the object of blame when things go wrong. The crisis of science's reproducibility—one thinks here of Ioannidis' (2005) warning bell, 'Why Most Published Research Findings Are False'—was laid at the door of statistics. This is not surprising, as statistics is the coalface of research: all sciences make use of statistical inference and most scientific results are couched in some form of statistical reasoning. This resulted in statistical wars (Mayo 2018, Stark and Saltelli 2018). Yet, few if any wondered how many model results were reproducible. If in the field of clinical medical research, the percentage of non-reproducible studies can reach 85% according to some estimates (Chalmers and Glasziou 2009), what is that in mathematical modelling (Saltelli 2019a)? You can pre-register a medical trial, but not a modelling exercise (see 'Pay No Attention to the Model behind the Curtain').

As noted by van Beek et al. (2022), models are apt at 'navigating the political': in the case of the all-important Integrated Assessment Models (IAMs), one can register a '"political calibration", a continuous relational readjustment between modelling and the policy community'.

Uncertainty quantification should be part and parcel of ordinary modelling work, as it is in statistics, where the confidence in the result is systematically the result of a test. Using statistics, an experiment may 'fail'. A modelling study rarely fails: as noted by Naomi Oreskes (2000), just quoted, it is impossible to 'falsify' a model—that is, to disprove it based on observations. To fail, a model should admit a situation of ignorance (see 'Mind the unknowns', and 'Models as metaphors'), but this would conflict with the PEMM.

Numbers and democratization: What about models?

Can a model be democratic? This may appear to be a weird question, since the quest for transparency demanded of models is a requisite for the exercise of democracy, and hence models—due to their capacity to inform policies—should in principle be democratic, or at least aspire to be.

Perhaps models can be democratic—and even revolutionary—when they are simple in their form. The model I = PAT, meaning that human impact on the environment (I) is a function, or simply the product, of population (P), affluence (A), and technology (T), was instrumental in warning about impending environmental catastrophes. The equation, put forward by ecologists in the 1970s, was in turn criticized by their opponents as a prophecy of doom that did not come to pass, never going out of fashion or ceasing to inspire debate (Chertow 2000). This may in turn illustrate well a point made by Pierre Duhem at the beginning of the last century in his objection to the nascent modelling enterprise: 'In Duhem's view models served only to confuse things, a theory was properly presented when cast in an orderly and logical manner using algebraic form' (Morgan and Morrison 1999). Clearly, the thesis that all models must equal the algebraic simplicity of 'IPAT' will not cut much ice in the world of petabyte computations. Yet models can be judged in light of the breadth or perspective they offer, as well as the level of debate they spur; we favour complexity in interpretation, rather than complexity in construction.

The vicissitudes of models and their predictions takes us to another point made by sociologists of quantification, and this is that—to oversimplify—numbers may be born liberators and end up dictators (Porter 1994, Power 2004). More accurately, one can say that the production of socially relevant numbers is 'part of a cycle of innovation, crisis and reform, which continually expands into new regions of social and economic life' (Power 2004), alternating between capturing and losing public trust in the process (Porter 1995).

For example, Didier (2020) recounts how the public availability of numbers about agricultural production helped farmers at the beginning of the twentieth century to defend against rough deals when selling their harvest. The great recession flagged a diminished relevance of these numbers for both government and farmers. Likewise, trust in and enthusiasm for mathematical models of increasing sophistication has been a recurring characteristic of economics ('Mind the consequences'), with Erik Reinert (2000) talking of cyclic tendencies of economics towards scholasticism in the pursuit of mathematical rigour.

Undeniably, numericization encourages economization, whereby governments and citizens are led to privilege objectives of efficiency and competition, of which the market is the best judge. As noted by Mennicken and Espeland (2019), these objectives 'run the risk of hollowing out democratic rule'. The same authors add that '[n]umbers, and their presumed transparency, both illuminate and obscure relations of power'.

These considerations come from sociology of quantification, which is mostly concerned with various indicators of public life, and how it is necessary—as shown by the statactivists (Bruno, Didier, and Vitale 2014)—for a literate public to make numbers part of public debate (Desrosières 1998). How about models? With models, the distance between numbers and lay citizens is greater, so citizens' literacy must be greater to be able to access model-generated numbers. Pursuing this direction is fraught with difficulties, not least that of falling into the deprecated 'deficit model' of the relation between science and policy, whereby the reason of public contestation to new technologies is not due to the non-desirability of the same technologies, but to the public not being literate enough to understand the science (Wynne 2014).

Mennicken and Espeland (2019) trace avenues for the investigation of the relation between numbers and democratization:

- Public participation and inclusion of local knowledge in (uncertain) indicator design, including classification, measurement, and aggregation
- The development of alternative measures—counter-quantifications
- The role of numbers in framing public discussion and deliberations about public goods
- The impact of new media and technologies on citizens' participation

How does this conjugate when the numbers come from models? In general, models appear to militate for their nature against the instances of democratization—they are more opaque, more easily associated with neutrality, and in fact quite effective in framing public debate and deliberation in a top-down tradition that runs from expert to citizen. For example, counter-modelizations are not impossible (see 'Mind the frames' and 'Models as metaphors') but surely more cumbersome to build and consume than the counter-statistics of statactivists.

As discussed in several chapters, we must therefore investigate what model quality means: is it the perfection of a mathematical model or its social responsibility?

It is necessary to adopt a spirit of constant questioning about this important trade-off. What is the real dilemma at stake? Once again, we would betray the spirit of the book if we answered this question definitively, although we do share the concern of Robert Salais (2022) and Amartya Sen (1990) for a double dimension of quality that is technical as well as normative.

Quantity or quality?

We have stressed in this work the importance of rules such as 'Do not sanctify the neutrality of the technique', 'Acknowledge that uncertainty is part of this complex world', and 'Respect docta ignorantia'—don't mislead the map for the territory, or the model for the analysis.

Our call to moderate quantification—whether from models, rankings, or other avenues—and to submit it to judgement, can be found in disparate quarters. From Winner's call (1989) not to blindly accept cost–benefit and risk analyses to the caveat of Muller (2018) that 'the best use of metrics may be not to use it at all', the caution is put forward by sociologists, ecologists, and practitioners. Schumacher (1973) makes the point with delicate care:

> [Q]uality is much more difficult to 'handle' than quantity, just as the exercise of judgment is a higher function than the ability to count and calculate. Quantitative differences can be more easily grasped and certainly more easily defined than qualitative differences: their concreteness is beguiling and gives them the appearance of scientific precision, even when this precision has been purchased by the suppression of vital differences of quality.

Yet these are minority views, as the temptation to traduce in numbers objects that are in essence fundamentally qualitative is deeply rooted in the existing cultures, notably in the practices of impact assessment (Saltelli et al. 2022, Stirling 2022).

Conclusion

Why should one accept the broad direction of framings offered by the present volume?

One possible answer is that our view is inclusive of voices and cases coming from different quarters, offering the occasion for reflection on the qualitative knowledge that one finds codified in a mathematical model.

Our authors shared their practices from different disciplines, from statistical modelling to composite indicators to economics and environmental sciences. Each domain has specific problems, and these were investigated, showing how boxes can be open from different sides: models for the COVID-19 pandemic have been looked at from different angles, including that of a jurist; models used in economics have been central to our discussion, but environmental studies were discussed as well.

Several authors have stressed the relative immunity of mathematical models from criticism, based on the epistemic authority conferred by mathematics and by the objective difficulty in understanding the fabric of a model. As a result—as we mentioned in the *Manifesto* that gave occasion for the present volume, models can easily be underexplored and overinterpreted.

The present volume also shows that when a model is immersed in the system for which it has been created, it can take on a life of its own, exerting a sort of gravitational pull that deforms the surrounding space. Another apt metaphor is that of a chameleon (Pfleiderer 2020), whereby it is at times unclear whether a model is a tool for studying a system or for producing actionable inference on the same system.

Our work also exemplifies the value of 'plural and conditional' expressions of knowledge. While this set of interpretations as a whole is plural, it is not indeterminate or all-encompassing. There are distinct strands of commonality amid the diversity of the voices that we shall try to summarize in these short conclusions.

The five rules we discussed in the volume allowed us to identify actors that should be involved in the reciprocal domestication between models and society. The 'solitude' of the model makers, who are the only ones able to interrogate a model and translate through a mathematical process a relevant segment of the reality they observe, appears to shield reason from passion, but hinders social negotiation.

The primacy of knowledge and disciplinary competence over qualitative analysis that is encouraged by models inevitably compresses the complexity of phenomena (see 'Mind the unknowns'), leading to obfuscation, reductionism, and invisibilities of issues or excluded actors. This inevitably raises a question about the quality of information underlying the model, including the trade-off between mathematical precision of a model and its ability to 'tell us' about a phenomenon that is not entirely measurable.

One point held by many in the present work is that models are performative: they do things, at the same time describing and creating, in a process that urgently needs more societal scrutiny. In other words, it appears crucial to adopt an assumption-hunting approach aimed at the views of developers, including possible rhetorical or instrumental uses. Such an approach could overcome the presumed neutrality of the model, avoid the model of everything, and not take for granted that the mathematical precision and the complexity of a model is a quality seal of the robustness of the model outputs.

One of our authors suggests that if we are not to throw out the modelling baby with the bathwater of modelling hubris, we need to make modelling more open, plural, and conditional. This process will be neither easy nor rapid, as entrenched cultures of quantification and numericization fostered by incumbent powers represent a considerable obstacle that becomes impregnable when disciplinary barriers get in the way. In a sense, most of the constructive solutions offered by our authors imply the removal of some disciplinary wall, allowing different gazes to peek above the shoulders of the number crunchers. We believe that statistics offers an example of how this can be done. Pointing modellers towards the way of statisticians—with their rich tradition of introspection—should be feasible without causing excessive offence to the rhetoric of evidence-based policy.

If this volume has achieved its objectives, it will have facilitated communication between modellers and sociologists of quantification, as well as between modelling and the general public.

Can we conclude that mathematical modelling is ineffective? No, we cannot. The examples discussed here confirm this: mathematical models produce lines of action, establish standards of optimizations, bind objectives, guide decisions, and impact on the reality they describe. So yes, mathematical modelling is effective.

If this effectiveness is not to transform into vulnerability, then models and society need more reciprocal attention and less automatism, more scrutiny and less pretension—in a word, better rules of coexistence.

References

van Beek, L., J. Oomen, M. Hajer, P. Pelzer, and D. van Vuuren. 2022. 'Navigating the Political: An Analysis of Political Calibration of Integrated Assessment Modelling in Light of the 1.5 °C Goal'. *Environmental Science & Policy*, 133, 193–202, doi: 10.1016/j.envsci.2022.03.024.

Borges, J.L. 1946. *Del Rigor en la Ciencia (On Exactitude in Science)*. https://ciudadseva.com/texto/del-rigor-en-la-ciencia (accessed 27 March 2023).

Bruno, I., E. Didier, and T. Vitale. 2014. 'Special Issue. Statistics and Activism'. *Participation and Conflict. The Open Journal of Sociopolitical Studies*, 7(2), 198–220.

Chalmers, I. and P. Glasziou. 2009. 'Avoidable Waste in the Production and Reporting of Research Evidence'. The Lancet, 374, 86–9, doi: 10.1016/S0140-6736(09)60329-9.

Chertow, M.R. 2000. 'The IPAT Equation and Its Variants'. *Journal of Industrial Ecology*, 4(4), 13–29, doi: 10.1162/10881980052541927.

Desrosières, A. 1998. *The Politics of Large Numbers: A History of Statistical Reasoning*, Harvard University Press, Cambridge, MA.

Didier, E. 2020. *America by the Numbers: Quantification, Democracy, and the Birth of National Statistics*, The MIT Press, Cambridge, MA.

Edwards, P.N. 1999. 'Global Climate Science, Uncertainty and Politics: Data-Laden Models, Model-Filtered Data'. *Science as Culture*, 8(4), 437–72.

Eker, S., E. Rovenskaya, M. Obersteiner, and S. Langan. 2018. 'Practice and Perspectives in the Validation of Resource Management Models'. *Nature Communications*, 9(1), 5359, doi: 10.1038/s41467-018-07811-9.

Funtowicz, S. and J.R. Ravetz. 1990. *Uncertainty and Quality in Science for Policy*, Kluwer, Dordrecht, doi: 10.1007/978-94-009-0621-1_3.

Gigerenzer, G. 2018. 'Statistical Rituals: The Replication Delusion and How We Got There'. *Advances in Methods and Practices in Psychological Science*, *1*(2), 198–218, doi: 10.1177/2515245918771329.

Ioannidis, J.P.A. 2005. 'Why Most Published Research Findings Are False'. *PLOS Medicine*, *2*(8), e124, doi: 10.1371/journal.pmed.0020124.

Jakeman, A.J., R.A. Letcher, and J.P. Norton. 2006. 'Ten Iterative Steps in Development and Evaluation of Environmental Models'. *Environmental Modelling & Software*, *21*(5), 602–14.

King, M. and A. Schüt. 1994. 'The Ambitious Modesty of Niklas Luhmann'. *Journal of Law and Society*, *21*(3), 261–87.

Louie, A.H. 2010. 'Robert Rosen's Anticipatory Systems'. *Foresight*, edited by R. Miller, *12*(3), 18–29, doi: 10.1108/14636681011049848.

Mayo, D.G. 2018. *Statistical Inference as Severe Testing. How to Get Beyond the Statistics Wars*, Cambridge University Press, Cambridge, doi: 10.1017/9781107286184.

Mennicken, A. and W.N. Espeland. 2019. 'What's New with Numbers? Sociological Approaches to the Study of Quantification'. *Annual Review of Sociology*, *45*(1), 223–45, doi: 10.1146/annurev-soc-073117-041343.

Morgan, M.S. and M. Morrison, eds. 1999. *Models as Mediators: Perspectives on Natural and Social Science*, Cambridge University Press, Cambridge; New York.

Muller, J.Z. 2018. *The Tyranny of Metrics*, Princeton University Press, Princeton, NJ.

Oreskes, N. 2000. 'Why Predict? Historical Perspectives on Prediction in Earth Sciences'. In *Prediction: Science, Decision Making, and the Future of Nature*, edited by D.R. Sarewitz, R.A. Pielke, Jr, and R. Byerly, Jr, Island Press, Washington, DC, pp. 23–40. http://trove.nla.gov.au/work/6815576?selectedversion=NBD21322691 (last accessed 26 March 2023).

Padilla, J.J., S.Y. Diallo, C.J. Lynch, and R. Gore. 2018. 'Observations on the Practice and Profession of Modeling and Simulation: A Survey Approach'. *SIMULATION*, *94*(6), 493–506, doi: 10.1177/0037549717737159.

Pfleiderer, P. 2020. 'Chameleons: The Misuse of Theoretical Models in Finance and Economics'. *Economica*, *87*(345), 81–107, doi: 10.1111/ecca.12295.

Pielke, R Jr. 2020. 'The Mudfight Over "Wild-Ass" Covid Numbers Is Pathological'. *Wired*, 22 April 2020 [Preprint]. https://www.wired.com/story/the-mudfight-over-wild-ass-covid-numbers-is-pathological (last accessed 20 March 2023).

Porter, T.M. 1994. 'Objectivity as Standardization: The Rhetoric of Impersonality in Measurement, Statistics, and Cost-Benefit Analysis'. In *Rethinking Objectivity*, edited by A. Megill, Duke University Press, Durham, NC, pp. 197–237.

Porter, T.M. 1995. *Trust in Numbers: The Pursuit of Objectivity in Science and Public Life*, Princeton University Press, Princeton, NJ.

Power, M. 2004. 'Counting, Control and Calculation: Reflections on Measuring and Management'. *Human Relations*, *57*(6), 765–83, doi: 10.1177/0018726704044955.

Ravetz, J.R. 1971. *Scientific Knowledge and Its Social Problems*, Oxford University Press, Oxford.

Reinert, E.S. 2000. 'Full Circle: Economics from Scholasticism through Innovation and Back into Mathematical Scholasticism'. *Journal of Economic Studies*, *27*(4/5), 364–76, doi: 10.1108/01443580010341862.

Rhodes, T. and K. Lancaster. 2020. 'Mathematical Models as Public Troubles in COVID-19 Infection Control: Following the Numbers'. *Health Sociology Review*, 29(2), 177–94. doi: 10.1080/14461242.2020.1764376.

Rosen, R. 1991. *Life Itself: A Comprehensive Inquiry into the Nature, Origin, and Fabrication of Life*, Complexity in Ecological Systems, Columbia University Press, Washington, DC.

Salais, R. 2022. '"La donnée n'est pas un donné": Statistics, Quantification and Democratic Choice'. In *The New Politics of Numbers: Utopia, Evidence and Democracy*, edited by A. Mennicken and R. Salais. Palgrave Macmillan, London, pp. 379–415.

Saltelli, A. 2019a. 'Should Statistics Rescue Mathematical Modelling?' *arXiv*, 1712(06457).

Saltelli, A. 2019b. 'Statistical versus Mathematical Modelling: A Short Comment'. *Nature Communications*, 10, 1–3, doi: 10.1038/s41467-019-11865-8.

Saltelli, A., M. Kuc-Czarnecka, S. Lo Piano, M. János Lőrincz, M. Olczyk, A. Puy, E. Reinert, S.T. Smith, and J. van der Sluijs. 2022. 'A New Gaze for Impact Assessment Practices in the European Union'. *SSRN Electronic Journal* [Preprint], (4157018), doi: 10.2139/ssrn.4157018.

Schumacher, E.F. 1973. *Small Is Beautiful: Economics as if People Mattered*, Harper Perennial, New York.

Sen, A. 1990. 'Justice: Means versus Freedoms'. *Philosophy & Public Affairs*, 19(2), 111–21.

Shapin, S. 2016. 'Invisible Science'. *The Hedgehog Review*, 18(3), 34–46.

Stark, P.B. and A. Saltelli. 2018. 'Cargo-Cult Statistics and Scientific Crisis'. *Significance*, 15(4), 40–3.

Stirling, A. 2022. 'Against Misleading Technocratic Precision in Research Evaluation and Wider Policy—A Response to Franzoni and Stephan (2023), "Uncertainty and Risk-Taking in Science"'. *Research Policy*, 52(3), 104709, doi: 10.1016/j.respol.2022.104709.

Weinberg, A. 1972. 'Science and Trans-Science'. *Minerva*, 10, 209–22.

Winner, L. 1989. *The Whale and the Reactor: A Search for Limits in an Age of High Technology*, University of Chicago Press, Chicago, IL.

Wynne, B. 2014. 'Further Disorientation in the Hall of Mirrors'. *Public Understanding of Science*, 23(1), 60–70.

Index

For the benefit of digital users, indexed terms that span two pages (e.g., 52–53) may, on occasion, appear on only one of those pages.

Tables, figures, and boxes are indicated by an italic *t, f,* and *b* following the para ID.